W0079759

Philosophy of Engineering and Technology

Volume 25

Editor-in-chief
Pieter E. Vermaas, Delft University of Technology, The Netherlands
General and overarching topics, design and analytic approaches

Editors
Christelle Didier, Lille Catholic University, France
Engineering ethics and science and technology studies
Craig Hanks, Texas State University, U.S.A.
Continental approaches, pragmatism, environmental philosophy, biotechnology
Byron Newberry, Baylor University, U.S.A.
Philosophy of engineering, engineering ethics and engineering education
Ibo van de Poel, Delft University of Technology, The Netherlands
Ethics of technology and engineering ethics

Editorial advisory board
Philip Brey, Twente University, the Netherlands
Louis Bucciarelli, Massachusetts Institute of Technology, U.S.A.
Michael Davis, Illinois Institute of Technology, U.S.A.
Paul Durbin, University of Delaware, U.S.A.
Andrew Feenberg, Simon Fraser University, Canada
Luciano Floridi, University of Hertfordshire & University of Oxford, U.K.
Jun Fudano, Kanazawa Institute of Technology, Japan
Sven Ove Hansson, Royal Institute of Technology, Sweden
Vincent F. Hendricks, University of Copenhagen, Denmark & Columbia University, U.S.A.
Don Ihde, Stony Brook University, U.S.A.
Billy V. Koen, University of Texas, U.S.A.
Peter Kroes, Delft University of Technology, the Netherlands
Sylvain Lavelle, ICAM-Polytechnicum, France
Michael Lynch, Cornell University, U.S.A.
Anthonie Meijers, Eindhoven University of Technology, the Netherlands
Sir Duncan Michael, Ove Arup Foundation, U.K.
Carl Mitcham, Colorado School of Mines, U.S.A.
Helen Nissenbaum, New York University, U.S.A.
Alfred Nordmann, Technische Universität Darmstadt, Germany
Joseph Pitt, Virginia Tech, U.S.A.
Daniel Sarewitz, Arizona State University, U.S.A.
Jon A. Schmidt, Burns & McDonnell, U.S.A.
Peter Simons, Trinity College Dublin, Ireland
Jeroen van den Hoven, Delft University of Technology, the Netherlands
John Weckert, Charles Sturt University, Australia

The Philosophy of Engineering and Technology book series provides the multifaceted and rapidly growing discipline of philosophy of technology with a central overarching and integrative platform. Specifically it publishes edited volumes and monographs in: the phenomenology, anthropology and socio-politics of technology and engineering the emergent fields of the ontology and epistemology of artifacts, design, knowledge bases, and instrumentation engineering ethics and the ethics of specific technologies ranging from nuclear technologies to the converging nano-, bio-, information and cognitive technologies written from philosophical and practitioners perspectives and authored by philosophers and practitioners. The series also welcomes proposals that bring these fields together or advance philosophy of engineering and technology in other integrative ways. Proposals should include: A short synopsis of the work or the introduction chapter. The proposed Table of Contents The CV of the lead author(s). If available: one sample chapter. We aim to make a first decision within 1 month of submission. In case of a positive first decision the work will be provisionally contracted: the final decision about publication will depend upon the result of the anonymous peer review of the complete manuscript. We aim to have the complete work peer-reviewed within 3 months of submission. The series discourages the submission of manuscripts that contain reprints of previous published material and/or manuscripts that are below 150 pages / 75,000 words. For inquiries and submission of proposals authors can contact the editor-in-chief Pieter Vermaas via: p.e.vermaas@tudelft.nl, or contact one of the associate editors.

More information about this series at http://www.springer.com/series/8657

Soraj Hongladarom

The Online Self

Externalism, Friendship and Games

 Springer

Soraj Hongladarom
Faculty of Arts, Department of Philosophy
Chulalongkorn University
Bangkok, Thailand

ISSN 1879-7202   ISSN 1879-7210   (electronic)
Philosophy of Engineering and Technology
ISBN 978-3-319-39073-4   ISBN 978-3-319-39075-8   (eBook)
DOI 10.1007/978-3-319-39075-8

Library of Congress Control Number: 2016943192

© Springer International Publishing Switzerland 2016
This work is subject to copyright. All rights are reserved by the Publisher, whether the whole or part of the material is concerned, specifically the rights of translation, reprinting, reuse of illustrations, recitation, broadcasting, reproduction on microfilms or in any other physical way, and transmission or information storage and retrieval, electronic adaptation, computer software, or by similar or dissimilar methodology now known or hereafter developed.
The use of general descriptive names, registered names, trademarks, service marks, etc. in this publication does not imply, even in the absence of a specific statement, that such names are exempt from the relevant protective laws and regulations and therefore free for general use.
The publisher, the authors and the editors are safe to assume that the advice and information in this book are believed to be true and accurate at the date of publication. Neither the publisher nor the authors or the editors give a warranty, express or implied, with respect to the material contained herein or for any errors or omissions that may have been made.

Printed on acid-free paper

This Springer imprint is published by Springer Nature
The registered company is Springer International Publishing AG Switzerland

Preface

The idea for this book grew out of the wonderful workshop on "Who Am I Online?" organized by Charlie Ess and Luciano Floridi back in May 10–11, 2010, at the beautiful Kalovig Center outside of Aarhus, Denmark. The idea behind the conference was to investigate the notion of personal identity as it applies to online self or online identity, precisely the topic of this book. Many scholars were invited to join the workshop. Apart from Charlie and Luciano, there were, as I remember, Stine Lomborg, Maria Bakardjieva, Wong Pak-Hang, Janice Richardson, Johanna Seibt, Dave Ward, Raffaele Rodogno, and many others. The idyllic atmosphere of the Kalovig Center was an ideal place for thinking together and engaging in common project of hashing out one's ideas in order to receive friendly feedback. I first conceived of the ideas presented in this book at the workshop. These ideas then developed and were further refined until they got their present shape in this book. This, however, by no means implies that the ideas are final. I don't know if there is any idea in philosophy that is final. Perhaps no philosophical idea ever is, and some philosophers do change their minds. But at least they represent what I believe to be the case and the book contains sustained arguments in their support.

The topic I presented at the Aarhus Workshop was "Who Am I Online? A View from Buddhism." In that I presented a straightforward Buddhist view on self and identity. This idea is by now quite well known so does not need to be repeated here. The argument I made then was that there is a correlation between the online and the offline worlds such that basically the same set of analytical tools can be applied in either. I still believe that this is the case. What I mean by the same set of tools is that, when we try to analyze and understand the situation of the "offline" self, that is, the self that all of us are familiar with, the tools, which also include the vocabulary and the theory that we use to describe and investigate the phenomenon, are the same no matter the self is there in the so-called "real" world or the so-called "virtual" world. Of course the self as existing in the latter world is the subject matter of this book. Here I say "so-called 'real' world" with a tongue in cheek. No one can deny that the world as we perceive it, in which we live and breathe, is not real, but I would like to point out that in today's world the real and the virtual are becoming more and more of the same substance. This does not mean that we are living in a virtual or simulated world, but I intend to mean that the two worlds are collapsing to each other and the boundary between the two worlds is not as hard and fast as many may believe (this will be more so when what is known as "ubiquitous" or "pervasive" computing becomes more common – I also investigate this phenomenon in the book). Thus, even if Buddhism was developed more than two millennia ago in order to analyze the "offline" self, the same analytical tools in Buddhism can also be used to analyze the "online" self too. This idea also underlies many views that are presented in this book.

However, I would like to point out that even if the book found its inspiration from the Buddhist perspective on the self, this is definitely not a book on the Buddhist view on the online self. That is, my plan is not to say that the self (whether offline or online) is of such and such characteristics because it says so in Buddhism. The plan is rather that I present a series of *independent* arguments intended to support the main theses of the book without relying on the authority of Buddhism. If Buddhist philosophy can be tenable and acceptable to the community of philosophers, it has to stand or fall on its own merit, not because this is what the Buddha taught or otherwise. In fact that would be contrary to the spirit of Buddhism too. Thus you will find the discussion on Buddhism forms only a small part of the book, so readers who are not Buddhists or who are not religious in any way can still benefit from the arguments presented here.

After the Aarhus Workshop I further developed the idea, resulting in a number of journal articles some of which are included in this volume. Naturally I am indebted to a large number of people without whom this book will not have been possible. First of all I would like to thank Charlie Ess and Luciano Floridi, the two co-hosts of the Aarhus Workshop, whose idea on having a meeting on "Who am I online?" sparked my interest in the metaphysics of the online self, a field that involves not only many branches of philosophy such as metaphysics, philosophy of technology, and ethics, but also many academic disciplines outside of philosophy as well, such as communication studies, sociology, anthropology, and history. So another benefit of the topic of this book is that it is interdisciplinary and is quite likely to attract interests of scholars in fields other than philosophy. Charlie Ess has been very helpful to me in many areas. Apart from being such a wonderful host during my Aarhus visit in 2010, our friendship actually developed well before that, dating way back to 1998 when he and Fay Sudweeks organized the first international conference on Cultural Attitudes toward Technology and Communication (CATaC), which has developed into a well-known series of conferences. I had the good fortune of being able to invite Charlie to Thailand twice and hope that our friendship and collaboration do continue. Luciano has been a constant friend who supports my attempts at presenting these philosophical reflections and gives me a generous number valuable comments and suggestions. I also hope that our collaboration continues.

I am also grateful to all the participants of the Aarhus Workshop whose challenges and criticisms of my presentation resulted in the development of the ideas found in this book. I would like also to thank Karamjit Gill, editor of the journal *AI & Society*, who invited me to contribute the paper on ubiquitous computing, and John Weckert, who has also been very helpful to me in various ways, one of which was that he invited me to contribute another of my paper to the online journal *Information*. Both papers play a large role in the development of ideas which led to this present book.

The road from the Aarhus Workshop to the book has been quite long. Along the way I am also fortunate to receive help and support from various people. Apart from the meeting in Aarhus, I also benefited from a meeting in Bangkok on "Online Studies," organized by the Thai Netizen Network in November, 2010. The informal and friendly meeting gave me a chance to present my work to people in other academic fields and for the lay public in Thai language. Arthit Suriyawongkul was as always a key person in the Thai Netizen Group who always gave me encouragement. My thanks also go to Elizabeth Buchanan and Michael Zimmer who invited me to talk in a keynote panel of the Computer Ethics/Philosophical Enquiry (CEPE) conference in Milwaukee, Wisconsin in 2011, giving me the opportunity to further reflect on the view that eventually found its home in this book. I would like to thank Philip Brey, Wong Pak-Hang, Johnny Søraker, Axel Gelfert, and Eric Kerr, all of whom play a role in one way or another in my philosophical development.

Bangkok, Thailand   Soraj Hongladarom

Contents

Chapter 1
Introduction

Today we find social media everywhere. In Thailand where I live and work, it is becoming more difficult to walk on the streets of Bangkok and find someone who is not looking at her smart phones, using her thumbs to chat with her friends and enjoying the pictures and texts offered there. On the new electric train system that is fast becoming a familiar scene in the city, people either sit or stand with their eyes focusing on their phones, ostensibly oblivious to what is going on around them. It is as if their brains are being plugged on to a giant network so that their reality is what happens on the screen of their phones rather than outside. In restaurants it is not uncommon to find couples sitting together at a table. In the old days they might look at each other's eyes and talking to each other, but today they look more at their own smart phones rather than at each other. The world is indeed changing. If these scenes are becoming a familiar sight in Bangkok, they are indeed happening everywhere. Admittedly these scenes are not happening outside of the Bangkok metropolis much, but it is getting there as smart phones are one of the top selling electronic items all over the country. People all over Thailand who can afford them snap them up very fast.

Most of the applications these people use when they look at their phones are the social media. Names are already familiar: Facebook and Twitter, and to a lesser extent but up and coming, Google Plus. In the recent past the names might include MySpace or Hi5, but these sites are largely neglected by the social media savvy users now. A trip back to Hi5, for a typical Thai internet user, recalls nostalgic scenes from the recent, pre-Facebook past. This is rather surprising because social media are in fact a new type of websites, having come on to the scene only a few years ago. Before these social media sites there were web sites that offered text-based discussion forums. These used to be highly popular in Thailand, as people found it exhilarating to be able to discuss almost anything, using almost any kind of language, with their peer from all around the country. This was something they were not able to do before. Certainly there were limits to this freedom. The draconian lèse majesté law against insulting the king is still in place, but even that could not dampen the enthusiasm and the speed with which Thai people gobble up web discussion

© Springer International Publishing Switzerland 2016
S. Hongladarom, *The Online Self*, Philosophy of Engineering and Technology 25,
DOI 10.1007/978-3-319-39075-8_1

forums and later social networking sites. Here in Thailand one popular website still survives and is in face thriving in the age of sophisticated social networking sites — Pantip.com. This shows a connection between the older discussion forum and today's social networking sites such as Facebook and Twitter. In fact many users have accounts in both Pantip.com and Facebook and they routinely share material from one to the other, thus merging the content of the two.

What is most interesting in the phenomenon of either the web discussion forum or social networking sites is how the users present their selves in the online world. In the former they did not have much leeway to do so, as the technology was not sophisticated enough. So they were limited to putting up their pseudonyms and perhaps a small picture to represent themselves. But in Facebook and Twitter these "online selves" are actually becoming more mature. Users have the ability to form their own profile page where they announce to the outside world who they are or, perhaps more accurately, who they want the world to perceive them to be. Thus this is the subject matter of this book. The main concern of the book is on the metaphysical constitution of the online self. What exactly is it? What are the relations between the online self and their "offline" counterpart, that is, the normal self with which we are already familiar? These questions are important because as people are more and more hooked up with social media, the role of the online selves is more and more visible and significant. Facebook advertises itself as a *social* media. That is, the website regards itself as a tool to connect people together. In order for that to be possible, people have to have an online presence on the website. In other words, they have to put up their online counterparts there. It is as if parts of their own selves are put there on the website. And when the social media play more prominent roles in people's lives, it appears as if the selves that are there play more roles, and seem to take on a more or less independent status by themselves. It is not uncommon nowadays to find people whose reputation depends more on their online selves than on their usual selves in the outside world. It is then an emerging phenomenon made possible by the social media, and since the latter is a product of information technology, another dimension of the problem to be investigated in the book is on how technology affects the sense of self and the relation between the self and the world. Social networking sites are made possible by technology; at one level of description the online self is nothing but a collection of ones and zeroes, as are all things digital. However, this collection of ones and zeroes can have very strong impact on the world, especially when people depend more on the social media and present themselves through these media. As the online self is becoming more real and as it plays more roles in lives, they have to be analyzed in terms of more than just ones and zeroes. And here is where the difficulty is. How could we best characterize the online self? What kind of language should be the one most suitable to describe it for our purposes? And as technology plays a constitutive role, what are the relations between the technologies of the internet and the online self?

Furthermore, as the online self plays more roles in society, it is bound to generate another sort of questions, one that concern ethical values. The idea behind social networking sites is that people represent who they really are in the online arena. That is, the offline and the online selves should match. But what happens if someone invents a new identity and presents it online in such a way that does not bear any resemblance to who he or she actually is? This is a big problem because the whole idea of social networking appears to depend on the authenticity of the two kinds of selves. What actually gets connected together on the social networking sites are, exactly speaking, online selves. I put my profile and my digital self online so that my digital self could connect with yours on the networking platform. But if there are discrepancies between the self I put online and my own real self, then how can my friends know that the one they are interacting with is indeed me? How can they find me on the Internet if I don't use my real name, for example, or don't advertise my pseudonym so widely that they can come to know?

We can see that there are a whole host of problems surrounding the online self, so much, I think, that this book could not cover all of them and can actually provide only a sketch of the possible problems. In any case, the main methodological concern of the book is philosophical. That is, I intend to use mainly philosophical tools and vocabulary to analyze and investigate the problems. This is not to say that empirical studies are not important. They are indeed very important, and I rely on them in cases where such studies help us understand more of the situation that is under discussion. Nonetheless, they are not the main concern. The main concern, on the contrary, is to look at the online self as a phenomenon for philosophical analysis. This importance of relying philosophical analysis can be seen most clearly, I think, when we analyze the role of technology in society. Here we can put the phenomenon of the online self and the social media in general as a topic under technology in society. Hence the book can also be regarded as a contribution in philosophy of technology. Philosophy of technology has been mostly concerned with how technology is related to human beings. Not in the sense of the humans as users or manipulators, but as beings who, phenomenologically, stand in relation to technology as the self stands to the other, or perhaps as one phenomenological being having a relation with another phenomenological being. In other words philosophy of technology is not concerned with humans merely interfacing with technology, but with the question of value that inevitably arises when humans stand face to face with technology.

1.1   Main Argument of the Book

The main argument I present in the book concerns implications of the view that the self is a composite entity and does not exist on its own. Many findings from neuroscience concur that the source of the sense of the self in an individual cannot be pinpointed in any particular region of the brain. In contrast to specific functions such as vision and hearing, self-consciousness is distributed globally throughout the brain and there is no one specific region that is responsible for it. Instead the sense of the self arises out of the awareness that one is a unified entity that is set against the world—the origin of the primordial sense of the subject and the object. In *The Ego Tunnel*, Thomas Metzinger argues in no uncertain terms that the self does not exist (Metzinger 2009). Instead the brain constructs a model of the self, what Metzinger calls the Phenomenal Self Model (Metzinger 2009, p. 2). In other words, the self is as much an illusion created by the brain, something those who watch the National Geographic series *Brain Games* will be immediately familiar. Metzinger refers to an experiment where the subject puts her right hand down under the table, leaving the other hand on. On the place of where the right hand should have been is a rubber hand instead. Metzinger shows that the subject somehow has an illusion that the rubber hand is her hand and feels something on the rubber hand as the hand is stroked by a feather. Of course she cannot actually feel anything because that hand is a rubber one and has no nervous connection to herself. Nonetheless her brain assumes that the rubber hand belongs to her self and starts to trigger the tickly feeling of being stroked by a feather (Metzinger 2009, p. 3–4). Here, then, is where a lot of confusion occurs when the issue of the existence of the self is discussed. On the one hand, scholars such as Metzinger argue that the self does not exist. This also goes along with the Buddhist tenet of the Non-Self. Basically the tenet is the same—the self does not, strictly speaking, exist. But the confusion starts when one tries to explicate what actually constitutes "strictly speaking." According to Buddhism, in normal, everyday conversation it would be absurd to maintain that the self does not exist. Even Buddhist philosophers have to refer to themselves from time to time. That is just a part of everyday language use. However, when analysis is applied, it is found, so Buddhist philosophy argues, that the self as normally understood is found to be nothing but a group of elements taken to relate to one another in one particular way. In other words, the self is in fact an illusion. But illusions do indeed exist, and in the rubber hand experiment the subject does indeed feel something when the rubber hand is stroked. So on the one hand the rubber hand does not belong to the subject's body—that much is obvious, but on the other hand the subject does feel something when the rubber hand is stroked, showing that her brain takes on the hand as a part of her body. According to the third-person objective model, the rubber hand is only a rubber hand, but according to the phenomenological, first-person viewpoint model, the rubber hand is part of the body. In the same vein, from one perspective the self does not exist (what exists are only blips inside the brain, for example), but from another the self clearly exist, as for example when one feels that *one* is having a headache. The dependence of the self's "existence" on perspectives also go along with what Buddhist philosophy has to say.

Antonio Damasio puts this point very well. Answering the question whether there is a self and if there is one, whether the self is present whenever we are conscious, he says:

> The answers are unequivocal. There is indeed a self, but it is a process, not a thing, and the process is present at all times when we are presumed to be conscious. We can consider the self process from two vantage points. One is the vantage point of an observer appreciating a *dynamic* object—the dynamic object constituted by certain workings of minds, certain traits of behavior, and a certain history of life. The other vantage point is that of the self as *knower*, the process that gives a focus to our experiences and eventually lets us reflect on those experiences (Damasio 2012, p. 8–9).

According to Damasio, the self operates at two levels, and we can safely say that what he actually means is the self-as-subject and the self-as-object, or in Damasio's terms the self as knower and as a dynamic object. The distinction between the self and other, subject and object, is very fundamental to the consciousness of a self; in fact the two constitute each other. We cannot have a consciousness of a self without a consciousness of the subject and object, and the other way round. Thus, it appears that there are two senses of the self. Viewed either as an object or as a knower, the self is a process in both cases. Its existence is a function of certain attitude that we take toward the world. In Buddhist term, this is what is meant by the expression "the self is not an inherently existing object." That is to say, the self does not exist on its own without being dependent on other factors.

In other words, the self is a second-order awareness; it arises when one reflects on the thoughts that one have, that one is having thoughts which are about things in the world, and *that* requires that there be someone who is the thinker of the thought. Instead of the organism only having a series of thoughts about the world around it which are distributed and diffused throughout the world, where the point of view of the onlooker can be simultaneously everywhere and nowhere, the organism has the perspective from a unified and specific location. This gives a unique perspective according to which the picture of the world comes to the organism as if it is the center of the world. Instead of having mere thoughts about the world which are nowhere and everywhere, the thoughts are those belonging to this unique perspective, thoughts that are related to a unique thinker, and this is what is meant by the first-person pronoun. What the recent findings in neuroscience shows is that this sense of a unique perspective is a second-order awareness arising out of many regions of the brain working together. This gives the self a unique position in that on the one hand, it very obviously exists, but on the other hand, since it is actually speaking a second-order perspective with which we arrange our perceptive awareness of reality, it does not exist within that reality. This gives rise to the famous Buddhist view that the self does not exist as an "inherently existing" entity. What the Buddhist means here is that the self does not exist at the first-order level of reality, the same level of rocks and flowers, but it certainly exists at the second-order level.

What this means for the online world is that we can compare thoughts about the world as postings and comments which are ubiquitous on the social network. Here I focus on the self as it appears in the online world, and does not look at how different forms of the online world might affect the character of the online self or not. In fact the self as it appears on, say, Facebook and Twitter are somewhat different one from the other. But that is not my main concern for the book. Without the sense of a self, those posts and comments belong to no one and they would be diffused to every corner of the online reality. Imagine posts and comments from no one in particular, those that can be found everywhere just as they are parts of the environment of the social network itself, then one gets an inkling of what it is like for thoughts without a thinker. In order for the posts and comments, whatever information is shared on the network, to belong to someone and for them to originate from a unified perspective, they have to come from someone and that someone is an online self. On

Facebook whatever I post and comment will be accompanied by my name (which could be my real name as is the case or could be any invented name) and my profile picture. This is the way Facebook groups all these posts and comments and shared pictures so that they always belong to *someone*. There are no posts or comments on Facebook which originate from no one in particular. However, those belonging-to-no-one posts are certainly possible on the network, and those are the correlates of belonging-to-no-one thoughts that we have seen earlier.

This view that the self arises out of second-order awareness forms the basis on which the philosophical analysis of the online self in this book is made. Another point that I would like to argue for is that the self is a composite entity. This is also old news. What I mean is that the term 'self' is more like a collective noun such as 'herd' or 'army' rather than one that denotes a single entity. A herd of cattle consist of at least more than one cow. Each cow in the herd, taken singly, is not the herd, but it is all the cows taken together that form the herd. In the same vein, a single soldier is not an army, but a number of soldiers joined together in a specific manner constitute an army. The self is a composite entity in the sense that there are a large number of elements that when taken together constitute a self, but when each of them is considered one by one none of them is a self on its own (in the same way as no one cow is a herd on its own). Thus, it is more how these elements are related together that is constitutive of a self rather than its internal characteristic (whatever that may be). In the online world, this translates to the situation where the online self is constituted by a number of posts, comments, status updates, shared pictures, etc. that can be grouped together as belonging to one particular online persona. Facebook tries to make that look easy by attaching a name and a profile picture on these postings, but that is actually tantamount to attaching a name on the various elements that together constitute a self. The problem is how exactly is it that these disparate posts and comments actually belong to a self in such a way that the self cannot be identified with any one of these postings, but to all of them taken together. My own writings and sayings are certainly quite numerous, and each of them can in principle have my name put on as a label, but that does not solve the problem of how these thoughts and sayings do constitute who I am as a unified self. It seems that the label comes later; it should not be the case that my unified self arises as a result of these thoughts and sayings are labeled as belonging to Soraj Hongladarom.

This issue of labeling points toward another important topic in this book, which is personal identity. In order to analyze what it means to be an online self, attention is paid to how an identity of the self or the person in the online world is established. The question has a long history. As we shall see in Chap. 3, Locke is credited as the first philosopher who takes up the topic of personal identity and proposes that one's memory is a necessary and sufficient condition for the continuity of one's own person through time. However, a serious problem with the memory account is presented even in Locke's lifetime by Bishop Butler, who argues that the memory account is circular because in order for me to be certain that the memory episode that I entertain is indeed what happened to me, there has to be an independent account that the story inside the memory indeed belongs to me, but according to Locke no such independent account is possible. Butler's argument here forms a basis of the argument I offer on personal identity, viz., the externalist account. Basically what the externalist position says is that what is going on inside one's subjective horizon is not sufficient to guarantee one's identity through time. One needs the help of external factors, i.e., what lies outside of one's subjective horizon. This can be anything, such as one's own birth certificate or the testimony of one's own mother. We will see this argument in more detail in Chap. 3. Translated to the online world, this means that an account of an identity of an online self or person also requires the help of external matter. It is not sufficient for one to be certain of one's own identity relying on one's memory alone. What ties up a post on the social network in, say, 2009, as an episode of my online self as exists now in 2014, needs some kind of verification from outside, such as coherence of the 2009 post with what happened then, and so on. This argument is also corroborated by recent findings in psychology that one's memory often fails and one sometimes mistakes a psychological episode as a part of one's own memory account.

The externalist account that I present in Chap. 3 leads to another major view that I argue for, namely the Extended Self View, which is the view that the location of the self can be extended outside of the brain and the body of the original owner of that particular self. I explicate this view in detail in the last section of Chap. 3. This view originates from Andy Clark and David Chalmers (Clark and Chalmers 1998; see also Menary 2010). It also resonates with Ciano Aydin's view of the mind as being artifactual in nature (Aydin 2013). Clark and Chalmers argue that, instead of regarding the mind as encased within one physical body (or one physical brain), the mind should be considered to extend outside of the body, such that tools that one use, for example, be an extension of one's own mind. Since what is there outside of the body and the skull is usually material, there is nothing in the extended mind thesis against the idea that the mind, extended outside, can be something material too. In this sense, then, smart phones are the quintessential extended mind. It is as if the brain is split open and comes in the form of a lovely rectangular object that one can hold and have a look at the beautiful images on it. As people depend more on their smart phones for cognitive tasks such as searching phone numbers, looking up facts, searching maps, performing calculation, setting up meeting schedules, and so on, it is as if these tasks, which we can see plainly as objective events, are instances of mental and cognitive tasks that are objectified in such a way that the smart phones are extensions of one's own mind. We shall also see this in detail in Chap. 3. As for the implication for the online world, the extended mind shows that one's mind, and hence one's sense of self, can be extended outward and thus can be extended to the cyberreality that is there in the social network or other types of online happenings too. In this case the online self is an extension of the real, offline self, and, as I shall also argue that there is a strong trend toward a merger of the online and offline worlds, the offline self is also an extension of the online one too.

My argument that the offline and online selves are extensions of each other appears to clash with the view that the self is intimately tied up with the physical body. Taking a cue from Merleau-Ponty's argument on the phenomenology of perception (Merleau-Ponty 1962), where the body is a necessary basis for perception and thus phenomenological account, many scholars and philosophers argue that the view that the self can be extended online is not tenable because the physical body, according to this view, is a necessary element of the self. However, the online self does not have to be completely divorced from the body. On the contrary, the Extended Mind Thesis explicitly allows that the mind can be extended toward material objects, only that those objects do not have to be one's physical, flesh and blood body. Furthermore, the externalist account of personal identity maintains that the external factor which is necessary for keeping personal identity can well be material, such as birth certificates, mementos from the past and so on. Thus the view that the online self is an extension of the offline one does not lead to the conclusion that the self is totally disembodied and divorced from all the ties with traditions and cultural roots. On the contrary, it is often those things from traditions and cultural roots that serve as external factors in maintaining personal identity and hence a sense of continued self as existing embodied in the community. So Julie Cohen's concern that the online self would be a disembodied and empty one, a kind of self that liberal philosophers assume to be the kind of self that is the subject of normativity, does not have to be realized (Cohen 2012). In her book, Cohen argues for a thicker kind of self, one which has embedded in it cultural, social and historical baggage that provides substantive content to the self. This is the core of Cohen's critique of liberalism, which is founded on the idea of an abstract self, the pure source of rationality and subjectivity. For Cohen, spatiality and the physical body necessarily mediates cognition and social arrangements (Cohen 2012, p. 40), so it might seem that the analysis of the online self offered in this book might fall afoul of her criticism. After all, the concern of the online self being overly theoretical or immaterial does not belong to Cohen alone. Scholars such as Dean Cocking and Steve Matthews also argue for the importance and necessity of the body in any attempt to analyze the online self or online identity (Cocking 2008; Matthews 2008). However, the account offered in this book does not fall afoul of these criticisms, since I do not argue that the online self is totally immaterial. In fact, as we shall see in the book, the online self is neither necessarily physical nor non-physical. That is, physicality or lack thereof is neither sufficient nor necessary for the online self. It is true that the online self primarily exists in the online world, the world that one finds primarily inside the computer monitor that one spends most of one's waking life watching. But it does not always have to be there. If the Extended Mind Thesis is tenable, what the mind can extend to does not have to be exclusively material or immaterial; moreover, the online self can have with it all the cultural, social and historical baggage that it likes as this baggage will form parts of the selves as the conception of the self extends toward them.

1.2 Structure of the Book

After the Introduction, Chap. 2 will relate the story of the self in Western and Eastern philosophies. The story will certainly be old news to those who have even a little background in history of Western philosophy, so they are advised to skip this part;

nonetheless, I attempt to tell the story of the self through its history in the philosophies of each tradition with an eye toward its relevance in helping us understand the online self better. Furthermore, a distinctive feature in this chapter is that I attempt to compare and contrast the notions of the self in Western, Chinese and Indian philosophies in the same chapter, something that I believe has not been done often at all. The advantage of putting together these three major intellectual traditions is that we can then see in one broad stroke how the self is understood and how aspects of the self are similar or different in these traditions, which will lead to a better understanding of the online self.

The story of the self in the West is familiar enough. It starts with the Greeks, whose conception of the self is radically different from the modern one that we are familiar with. In short, the Greeks do not separate between the mind and the matter, or the subject and the object, as we moderns do. On the contrary, they regard the mind as more or less material, as evidenced in Aristotle's conception of the self and the soul. In other words, the Greeks do not put a strict separation between the mental and the physical that we, inheriting from Descartes, tend to do. Minds are also material, and matter is also mental, as we can see from the Aristotelian conception of the final cause, where matter *aspires* to arrive at its final resting place according to its own nature. Then the story comes to Descartes and Spinoza. I spend relatively more time on Spinoza because his view on the self and the mind do not receive the attention that it deserves. According to Spinoza, mind and matter are aspects of one and the same reality. An individual mind, that which belongs to an individual person, is somehow a copy image, or a perfect correlate of the person considered as a physical body. The individual mind is then the "idea" of the individual body, where the idea and the body are attributes of one and the same segment of the one substance that constitutes the totality of what there is. This view has a strong relevance in our attempt to understand the online self because it helps us see that the online self can be both physical and mental at the same time. This is, it seems to me, another way of putting Clark's and Chalmers's thesis of the extended mind. What the mind is usually extended to is material—a notebook, a set of index cards, an iPhone, for example—but Spinoza's view of the ultimate identity of the mental and the physical helps us see this connection and puts it in a philosophically systematic manner. Then the chapter discusses Locke's view on the mind and the self. Locke's view will be more prominent in Chap. 3 when his famous view on the memory account of personal identity is discussed. Then Kant's important view on the self is also discussed extensively. What Locke and Kant share is a view of the self as an autonomous individual subject, a member of the Kingdom of Ends, in Kant's way of putting it. This is a linchpin of the liberal theory of normativity that we find in liberal moral political philosophers such as Rawls and Scanlon (See Rawls 1971; Scanlon 1998). What I argue is that the liberal conception, which is based on the Lockean and Kantian view of the self, does not do the fullest justice on the phenomenon of the online self, which I believe is better illuminated through the lens of the Spinozistic and Hegelian perspectives. Both Spinoza and Hegel maintain that the self is not separable from its material basis, and that it is not autonomous and fully individualized as a self-subsisting entity. This view of the self as more permeable and flexible makes us see better how the online and offline selves are interrelated. The online self is not a mere offshoot of the offline self, but both the offline and the online selves co-constitute each other; they both are integral parts of one and the same self, the self of an individual person. However, what sets an individual person apart from the world and from other persons also depend external factors, and not on the internal structure of the individual himself.

After discussing the history of the self in the West, I then go on to discuss the self through the eyes of the two main traditions of Asian philosophies, i.e., Chinese and Indian. Since there is a lot of material to be covered in the chapter, I focus only on Confucianism and Buddhism as representative of the Chinese and Indian traditions respectively. Certainly these are by no means truly representative, as there are many other strands in either tradition that are widely different from Confucianism and Buddhism, but due to the constraints imposed by the size of the book we have to limit ourselves to these two main traditions. According to Confucianism, the self is relational in that what it means to be myself depends on my roles and relations that I have toward my environment. Thus in order for me to be myself, I am also a professor (in relation to my students), a father (in relation to my son), a friend (to my friends), a husband (to my wife), a son (to my mother), and so on. This view is not the same as the Hegelian view in the West, which appears to be also quite relational. The difference is the Confucianism is clearly anti-metaphysical. It is not the concern of Confucian to answer the questions that beset Western philosophers, such as where we come from, what the world is made of, and so on. Instead the concern of the Confucian philosophers is the here and now—how to arrange a perfectly ordered community and state, how the family form the basis of justice, and so on. A contribution of the Confucian perspective in analyzing the online self is that the latter would then be regarded as a node in a web of relations, deriving its identity and being from the web rather than from its internal structure.

Then we come to Buddhism. Buddhist philosophy famously has a lot to say about the self, basically that it does not exist. This startling argument has given rise to a countless number of scholarly works attempting to debunk, analyze, explicate and support it. There are, fortunately, a number of good scholarly works that explicate this difficult topic (these can be found in the next chapter). What I do in Chap. 2 is only to provide a briefest sketch of the Buddhist Theory of Non-Self. Basically speaking, the Buddhist does not assert *tout court* that the self does not exist. That would be absurd. Instead the Buddhist tries to argue that the self as we experience it every day is an illusion. An often used example is a rope, which one mistakes to be a snake. The illusion disappears when one examines the rope more closely and finds out that it is not a snake after all. At first one sees what appears to be a snake, believing that that is a snake. This is analogous to the situation where one mistakes something else to be a self, such as one's own self. However, on closer inspection one finds that what one believes to be one's self in fact is not so, but something else. This is the essence of the Buddhist argument on the Non-Self. Another way of putting this is to compare the self to an army or a herd of cattle as we have done before in the chapter. Thus at one level of discourse the talk of the self is still necessary, but at another level, such as the metaphysical one where one tries to arrive at a more fundamental view of what there is, the self disappears and replaced by a succession of episodes, both mental and physical. This is not the same as arguing that the self is reducible to these episodes, but it means that the self *does* exist at one level of perception or analysis and disappears at another. We certainly discuss this important topic in more detail in the next chapter.

In Chap. 3, the aim is to sketch and explicate the thorny debate on personal identity. Here is where I present the main argument of the book, namely the Externalist Account of the personal identity problem and the Extended Self View. The traditional philosophical debate on personal identity is relevant to the analysis of the online self in several ways. Firstly, in order to understand how the online self derives its unity and identity through time, a thorough understanding of personal identity is crucial. Traditional arguments on the topic, ranging from Locke's memory view to the latest one in the literature, such as Olson's and Schechtman's, all can shed light on the situation of the identity of the online self. Moreover, analysis of the online self also, or so I argue, illuminates and enlivens the debate on personal identity, as there is an added dimension to the debate where what constitutes a person does not always have to be a physical body any longer. As mentioned earlier in this chapter, the Externalist Account takes its cue from Bishop Butler's criticism of Locke, who argues that one's autobiographical memory is necessary and sufficient for maintaining one's own identity. The chapter also discusses Derek Parfit's famous argument that psychological continuity is necessary and sufficient for personal identity, as well as Olson's Bodily or Biological Account and Schechtman's Narrative Account. All three, Parfit's, Olson's and Schechtman's, are found wanting because they are all internalist in the sense that they rely on factors *internal* to the subjective horizon of the person in question, thus cannot avoid question begging when it comes to establishing the identity of the very same person. Then I argue that the Externalist Account is the best way of coming to terms with the online self, as the identity of the latter is best construed as constituted by external factors. What I have in mind is that externalism in personal identity is analogous to externalism in epistemology. Recall that in epistemology, externalism is the position that seeks justification of belief outside of the subject's internal horizon, so to speak. The hitherto dominant Cartesian view holds that whatever that could justify a belief must be fully accessible to the subject, such as when I want to ascertain that the rose before me is red, I have to rely on the sense data that I am having while I here now am looking at the rose as the source of justification of my occurrent belief that the rose before me here is red. Justification is always internal. However, many epistemologists such as Goldman and others argue that facts of the world are what provide beliefs with justification such that they can become knowledge. In the same vein, my argument is that, instead of searching for the criterion of personal identity from within the conceptual and perception horizon of the individual, what ascertains identity of the person has also to be found outside, such as in the environment where the subject grows up. The relevance of this on the analysis of the online self is that the online self is best analyzed through the Externalist Account. That is, the identity of an online self is maintained through the relations that that self has toward other factors surrounding it. For example, we believe that a status update from someone in 2009, say, belongs to the same person as a comment made in 2014, not through the use of the same name and portrait (though these help a lot in normal cases), but through the network that that person (i.e., the one who posts in 2009 and the one who posts in 2014) makes throughout his presence in the online world. An illustration of the network is the group of friends that the person has in 2009 compared with the friends the person has in 2014. If these groups contain largely or mostly the same people, then a case can be made that the two persons are the same. Other factors play a role in this too. To be sure, the group of friends obviously is neither sufficient nor necessary in maintaining the person's identity, but according to the argument I put forward, it is very difficult to find really sufficient and necessary conditions for personal identity, since this is a fluid concept. What we can do is to try to establish until we are satisfied that one person in the past is the same as the person we do encounter at the moment. This is true in the offline world, and it's true in the online world too.

In the section on the informational self and the role of the body, the main idea is that I largely follow Luciano Floridi's view that the self is constituted through information. And by extension the online self is also constituted through information (which is easier to understand because the online self as appears on the computer screen is nothing but a collection of pixels processed by the data processing unit inside the computer). However, this does not mean that the self, either in its online or offline incarnations, is totally abstract, because the fact that things are constituted by information does not imply that they do not have mass or are unextended. On the contrary the mass and the extensional dimension of the thing are only possible because they are specified through information. The mass is usually expressed in numbers, so are the dimensions (height, length, and width). The new technology of 3D printing also illustrates that it is possible for extensional object to be constituted through information, as the information is used to specify the dimensions and other parameters of the object so that its 3D printing is possible. Furthermore, I discuss the role of the Extended Mind Thesis and the online self in the next section. The idea is that, as the mind can be extended outside of the body, it can certainly extend to the cyberreality of the online world; hence the online and offline selves are extensions of each other, as I said earlier. The online and offline selves in this case then jointly constitute the self of the person who participates in the online world and who put up his presence via an avatar or a social networking profile. The information making up his social media presence is every bit equal to the information making up his bodily presence as both contribute to making up the total of who he is, i.e., his own self and identity. Then the chapter concludes with a section specifically on the Extended Self View and how it differs from Chalmers and Clark.

Chapter 4 is an attempt to situate the analysis of the online self in the context of discussions in philosophy of technology, especially as regards to its critical aspect, which attempts to analyze technological phenomenon in terms of its impact mainly on socio-economic conditions. I discuss the views of some leading philosophers of technology, namely Heidegger, Marcuse, Borgmann, Ihde, Dreyfus, and Feenberg, especially concerning the online self and identity. What distinguishes the views of these philosophers is that they are more critical than the mainstream analytic philosophers such as Parfit or Olson, who are more concerned with conceptual analysis. Heidegger, Marcuse, and others discussed in this chapter try to see deeper than conceptual analysis and to come up with conclusions that have implications for policy implementation and instigating concrete changes in society. Even though Heidegger and Marcuse died long before they could see the online self or the Internet or the social media, their analyses were prescient and can readily be adapted to shed insights to the situation of the online self and the social media in general. In discussing these giants in philosophy of technology, the chapter can be regarded as a brief survey of the main thoughts on the subject. Those already well versed in Heidegger, Marcuse or Ellul can skip these sections and go right in to the section where I present my original contribution, which are about agency and continuity. The first topic deals with the possibility of the online self as an agent. After all, if the online self is to be a self, it has to have some capability of being an agent, for what is a self if it has no agency capability? Here I argue that the fact that an online self can be an agent is not mysterious considering that the online self is an extension of the offline one. We can perhaps understand this better if we take the online self to be a persona, a mask, worn by the offline self who has the agency and subjectivity required. But to view the online self merely as a mask worn by the offline self does not seem to do justice for the former since it has become increasingly highly complex in an environment where the social media platform has taken more and more visibility as an ersatz reality. In any case, if the two selves constitute each other according to the Extended Mind Thesis and the Externalist Account, then to view the online self as a mere persona would seem to be inadequate. Furthermore, Floridi's interesting distinction between responsibility and accountability is also discussed. For Floridi a rescue dog can be held accountable for what he has done to save people's lives; thus such dogs are often commended publicly for their bravery. However, no one currently holds that the dog is responsible for his action. Floridi uses the dog analogy to argue that a robot can also be accountable for its action (as again in rescue attempts) while it is certainly not responsible. In the chapter I argue that the dog analogy might not work as well as Floridi would have liked for robots or autonomous software programs. As for the problem of online agency—who is responsible for the action done by an online person?—the answer I give is that the person behind the online self is certainly responsible. Even if the software is autonomous to a certain extent and executes action somewhat independently, still the person who is responsible for putting the software into operation should bear some responsibility. As long as the software is not fully conscious, the burden of finding who is responsible and legally liable should there be something unlawful should fall on those who are responsible for all this in the first place.

Chapter 5 is to me one of the most interesting of all the chapters in the book. One important characteristic of social media is that they enable networking of "friends." However, it is unclear what these "friends" are on the social networking sites. Facebook makes it necessarily the case that everyone who is connected within my network is my "friends" even though some of these "friends" are those whom I have never met in real life, and I don't even know the real names of many with whom I connect. This is because a rather common practice in Thailand is that people tend to hide their real identities when they get connected to the social media. This defeats the original purpose of Facebook, to be sure, but for these Thai people they do not want to disclose their real identities because they do not find them necessary in engaging with others on the social networking. When they get on Facebook, they usually use an invented name and the profile portrait looks like anything but themselves. There are already a number of empirical studies on this topic, but for our purpose in this chapter the aim is more philosophical. It is to go back to what philosophers said about friendship in general and to find out how these ancient insights play out in the contemporary situation of online friendship. When people get to become "friends" who do not even know their real names, what is actually happening? Does this mean that the meaning of "friends" is being stretched almost beyond recognition? I begin the chapter with a historical narrative of what Aristotle has said on the topic. As is well known Aristotle is one of the few philosophers who said anything substantive about friendship, and his account in the *Nicomachean Ethics* (Aristotle 1962) has remained standard for all subsequent analyses of friendship until today. Moreover, I also discuss the Buddhist perspective on friendship, a topic which is much neglected in the literature, not only in philosophy and communication studies, but also in Buddhist studies. While there are some minor differences between the Aristotelian and Buddhist conceptions, I show that the similarities are substantive as both insist that friendship is necessary for the fulfillment of virtuous lives. The views of Kierkegaard and Nietzsche, two of the leading Western philosophers whose views on friendship are very interesting, are also presented. I show in the chapter that both of their views are relevant to the discussion of online friendship in significant ways. Nietzsche views genuine friendship as one where contest necessarily plays a role. That is, there has to be certain differences in genuine friends such that they contest each other in a common search for truth and understanding. Kierkegaard, on the other hand, views genuine friendship as one that exemplifies the ideal Christian love that does not distinguish one human being from another. These two contrasting views on friendship, I shall try to show in the chapter, can be found on the online platform and thus online friendship is capable of exemplifying either Kierkegaard's or Nietzsche's view. The upshot then is that being online or not is not relevant to being genuine, virtuous friends.

Basically my argument is that online friendship does not have to be estranged from the Aristotelian conception of ideal friendship, or Nietzsche's or Kierkegaard's for that matter. This does not mean that online friendship is automatically virtuous and ideal; there are other necessary factors, but I contend that there is nothing in online friendship that prevents it from achieving the ideal status outlined in Aristotle. Cocking and Matthews (2000) argue that text-based friendship is limited because text-based communication can only present only one side of the identity and personality of the friends. This makes it easy for the users to control how they appear in the eyes of their friends; hence for Cocking and Matthews the friendship is not genuine. On the contrary, I argue that even in text-based communication real friendship can occur, as in the past when people became friends and remained genuine friends only through correspondence.

The last chapter is about the online self in computer games. The self appears most often as the avatar, a Sanskrit term meaning "to come down." This is referred to a god, in most cases the god Vishnu, Preserver of the Cosmos, coming down and taking human or animal forms in order to fight against the evil that is bent on destroying the cosmic order. A key issue here is the relation between the game player and her avatar. Is the avatar mere "cursor" that responds to the command of the user in her navigating the terrain in the game and nothing more? Or does the avatar take a life of her own, so to speak, when she participates in the world of the game? Rune Klevjer (2012) argues for a "prosthetic telepresence" of the game player onto the game world via the avatar. His view is that the player is immersed into the game world during the play as if she herself were in the game. However, my difference from Klevjer in this regard is that, according to the Extended Self View that I advocate, the situation is better described as there being two selves, one belonging to the player in this world, and the other belonging to the avatar in the game world. An advantage of this way of looking is that one still maintains one's presence in one world while one is also present in the other. When one is "tele-ported" to the other world, one disappears from this world in order to appear again in the other—it is not the case that one can have two appearances at the same time. However, according to the Extended Self View, this is possible, and in fact this is a situation that is familiar to all game players. On the one hand they have their own presence in the actual world, while at the same time they have their own presence in the context of the game world. This is possible not only because the elements comprising the two selves actually do comprise the bigger self of the player plus her avatar (considered now as one big unit consisting of two smaller selves), but also because the game world itself is continuous with the actual (non-game) world. It is continuous because the computer game operates in the real world condition, requiring a processor, a lot of electric power, and so on. Thus when the player sends out an avatar, the two in a certain sense do actually exist side by side in the same world. The story of the movie Avatar, where the paraplegic soldier plays a real-life, walking and kicking avatar of an alien intelligent life form, is an illustration of this is possible. If the walking and kicking avatars (which can actually kick someone causing real pain) are there just to play games, then we can see how the context of games and avatars can merge with what goes on in the real world. Later in the chapter, I discuss the recent development of brain-to-brain integration technology, where information is directly shared among brains through electric wires or some other medium. This points to the possibility of merging two or more brains together, creating a possibility of enlarging brains, or, more accurately, of creating a network among brains, thus enhancing their capabilities than would be possible if each brain were to work alone. I discuss the implications of this technology for computer games and for understanding the relation of avatars and the selves of the game players. Then the book ends with a brief concluding chapter.

References

Aristotle. (1962). *Nichomachean ethics*. (M. Ostwald, Trans.). Indianapolis: Bobbs Merrill.

Aydin, C. (2013). The artifactual mind: Overcoming the 'inside–outside' dualism in the extended mind thesis and recognizing the technological dimension of cognition. *Phenomenology and the Cognitive Sciences, 14*(1), 73–94.

Clark, A., & Chalmers, D. J. (1998). The extended mind. *Analysis, 58*, 7–19.

Cocking, D. (2008). Plural selves and relational identity: Intimacy and privacy online. In J. van den Hoven & J. Weckerts (Eds.), *Information technology and moral philosophy* (pp. 123–141). Cambridge/New York: Cambridge University Press.

Cocking, D., & Matthews, S. (2000). Unreal friends. *Ethics and Information Technology, 2*, 223–231.

Cohen, J. (2012). *Configuring the networked self: Law, code, and the play of everyday practice*. New Haven: Yale University Press.

Damasio, A. (2012). *Self comes to mind: Constructing the conscious brain*. New York: Vintage.

Klevjer, R. (2012). Enter the avatar: The phenomenology of prosthetic telepresence in computer games. In J. R. Sageng, H. Fossheim, & T. M. Larsen (Eds.), *The philosophy of computer games* (pp. 17–38). Springer.

Matthews, S. (2008). Identity and information technology. In J. van den Hoven & J. Weckerts (Eds.), *Information technology and moral philosophy* (pp. 142–160). Cambridge/New York: Cambridge University Press.

Menary, R. (Ed.). (2010). *The extended mind*. Cambridge, MA: MIT Press/Bradford.

Merleau—Ponty, M. (1962). *Phenomenology of perception*. (S. Colin, Trans.). London: Routledge.

Metzinger, T. (2009). *The ego tunnel: The science of the mind and the myth of the self*. New York: Basic Books.

Rawls, J. (1971). *A theory of justice*. Rev. Ed. Cambridge, MA: Harvard University Press.

Scanlon, T. (1998). *What we owe to each other*. Cambridge, MA: Harvard University Press.

Chapter 2
The Self Through History

2.1 Introduction

Before coming to the online world, the self has had a long history and become subject of many different philosophical and religious traditions. In this chapter we
review a brief history of the self, mainly through how it is viewed by various important philosophical and religious traditions. It is hoped that by having a look at the
history, we gain a clearer picture and understanding of the characteristic of the self
as it is presented online. Furthermore, we also hope that the picture will reveal
aspects of the self in the offline world with which we are perhaps more familiar. It
will also help to show that there are no essential differences between the two kinds
of self.

In any case, we will present two major strands of thought about the self, one from
the West and the other from the Eastern tradition. This is surely a very large undertaking, one that cannot be attempted in any detail in a book, let alone a chapter such
as this one. There are in fact many fine histories of the self, both in the West and in
the East (such as Taylor 1989; Martin and Barresi 2006; Siderits et al. 2011; Ganeri
2012). Our purpose here is not to cram those histories into the limited space of this
chapter, but to paint a broad picture of how the self comes to be and how they are
viewed in the West and the East. (A book that tells the story of the self in the West
is Martin and Barresi 2006, but that only tells the Western part of the story, not both
the West and the East as this chapter attempts to do.) It is commonly known, however, that the West and the East are little more than catch-all terms which contains
many diverging elements inside, and those elements often strongly contradict one
another and argue against one another vehemently. For example, it might be thought
that the West has an essentialist conception of the self, where the self is understood
as something that exists independently having its own metaphysical status in such a
way that its being does not depend on any other entities. A good way to illustrate
this view is to compare this kind of self with the pit inside a peach. It is hard and it
is patently there. On the other hand, Eastern conception, according to this conception,

© Springer International Publishing Switzerland 2016

S. Hongladarom, *The Online Self*, Philosophy of Engineering and Technology 25,
DOI 10.1007/978-3-319-39075-8_2

tends to look at the self in much more relational terms. Instead of looking at the self as something existing independently, the Buddhist conception, so the common perception goes, looks at the self as emerging from the various relations that the self bears with other external elements, such as other selves and other entities. Without those interrelations, there is no self, and a metaphor would be a knot consisting of many strands of thread: The knot would not have been in existence were it not for the strands themselves. However, I would like to show in the course of this chapter that this picture is rather an over simplification. There exists elements in both the Western and Eastern thought that lend themselves to this kind of oversimplified interpretation, but as each tradition contains within itself a very large and diverse set of thoughts many of which argue against one another. In the West the idea that the self is more of a knot than a pit can be found in the thought of many modern philosophers such as Nietzsche and the postmodernist philosophers whose task is to decenter the subject. In the East, there are strong philosophical and religious traditions that maintain that the self is the core of identity which exists independently of any other external elements in almost exactly the same way as in the typical Western conception. Various Indian philosophies maintain that the soul is indestructible and has an independent metaphysical status. This conception stands in conflict with the common perception mentioned earlier.

Thus, instead of bifurcating the strands of the history of the self into Western and Eastern, what I will do in this chapter is to treat these strands as if they belong to the same large strand which is the whole of human philosophical attempt to understand the self. In other words, the purpose is not to do the usual task of comparing and contrasting what is there in both traditions, but to paint a broad picture of the history of the self using resources from both the East and the West. In what follows, however, I will discuss each of these main strands in turn, starting with the West, but this should not be taken to mean that the strands are separate, nor should it be taken to mean that there are distinct lines of thought that are unique in either tradition. The separation of the following headings into "West" and "East" is done only for the sake of convenience and organization.

2.2   The Self in Greek Philosophy

Ancient Greek philosophy does not have much to say about the self until Socrates. There was a momentous shift when Socrates turned away from speculation about the material nature that preoccupied those philosophers before him and became interested in human affair. Philosophers such as Thales or Anaximander were interested in finding the substance of material reality. It is well known that for Thales he thought water was the underlying substance of all natural phenomena and Anaximander thought it to be something indefinable and limitless, or *apeiron* in Greek. Either way the idea is that these philosophers look at the natural world and form questions about their ultimate origin and nature. It was not until Socrates and the Sophists who turn the attention away from nature and focus instead on what we today call problems in ethics and social philosophy. Instead of the question what ultimate composition of the world looks like, the Sophists and Socrates are interested in questions such as "What is virtue?", "What is the good life?", "How can I myself achieve the status of blessedness or happiness?" It can be seen that these questions are all centered on the self. When one asks what virtue really is, one is not interested in knowing about virtue in itself, but in relation to what one can do in order to achieve one's aim. Furthermore, the discussion of one's aims naturally involve comparing different aims and find a theory which would tell them what is the best aim that a human being should aspire to, the best kind of living that a human can possibly achieve. All are centered on the individual and the human being and how they live. This is in contrast to the earlier pre-Socratic philosophers who concern themselves primarily with the natural world. This, however, is not to say that these earlier philosophers are not interested in human affairs at all. Philosophers such as Heraclitus and Parmenides could be seen also to concern themselves with these questions; for example, they view that human blessedness could be achieved through realization of the ultimate nature of the cosmos—for Heraclitus it is the realization that all things change and for Parmenides it is that nothing changes. Blessedness for these philosophers could be achieved through aligning oneself along the ultimate nature of reality. However, the writings of Heraclitus and Parmenides that come down to us are too scant for us to say this with confidence, so this is only an interpretation, or perhaps an educated guess. In any case, it is clear that Heraclitus and Parmenides appear to break away from the concern of the earlier pre-Socratics who were interested only in natural phenomena. In saying that all things change or all things are one, both Heraclitus and Parmenides appear to break away from the concern of the earlier pre-Socratics in that they look at the natural phenomena not in order to know their natural cause and underlying substance, but their more ultimate nature in metaphysical terms.

In any case, it should be safe to say that the self emerges in its full-fledged form in the thoughts of Socrates and the Sophists. The dispute between the two is centered on what to do and how to understand the self. One of the most important arguments between Socrates and the Sophists concerns Protagoras' famous saying: Man is the measure of all things. Here we see clearly that the self does emerge as an entity that is the subject of philosophical dispute in its own right. The question is whether the speculations and discussions on the nature of reality that have preoccupied philosophers in the times leading up to Socrates and the Sophists are essentially "man-made" or is there a way for these discussions to lead to the truth which is objective and is free from human construction? In other words, when philosophers argue among themselves about whether water is the ultimate substance or air or some indefinite substance, are these questions merely *human* speculations and argumentations with no way to get at the real truth, or are they real questions in the sense that there is a real way to get an the objectively right or wrong answer? Are human beings at the center of the universe, free to create anything for their own use, or are they just one element among many, embedded within an objectively existing entity which they do not create, but one which they have a means to know exactly?

Even though they pioneer philosophical discussions on human agency, neither Socrates nor the Sophists has any substantive theory on the self. The task falls to Socrates' student Plato, who had an elaborate theory about the self and the soul. In fact the Greeks traditionally believe that the soul transmigrates after the body is dead, something that is similar to many strands in Eastern thought. Plato has a theory of conflicting pulls on the self; he compares these pulls to horses that pull the mind or the self into different directions. It is the self's duty to realize what is best for it, which is to return to the realm of the eternal Forms. The soul or the self has to fight off pulls from the carnal pleasures, which try to drag it down to earth. The Highest Good is the realm of the Forms; the soul has the capacity to know this and naturally tries to find its way back to where it belongs. Here one encounters the doctrine of *akrasia*, which is attributed to Socrates himself. According to Socrates, it is not possible for the soul or the self to know the Truth and act in such a way as to alienate itself from it at the same time.

The picture that emerges is that of the self and the soul as immaterial and existing independently of any other thing. Socrates' *akrasia* principle shows that the soul's knowledge is capable of controlling its behavior. The soul has, in other words, complete control over the body and its desires. When it is not possible for the soul or the self to knows intimately that something is bad and yet succumb to the bodily desire to do exactly that bad thing, the implication is that the soul and the power of rational thought has total power over the body. The *akrasia* principle has generated a very large number of discussions and debates over the centuries, but our purpose here is to depict the soul as it is understood by the Greeks, and we see that the picture that is emerging from Socrates and Plato is that of the soul as immaterial and capable of existing independently of anything else. The soul is also immortal, as can be seen from many passages in Plato's writings where the soul transmigrate and used to dwell in the realm of the Eternal Form before falling down to earth but always yearning to go back to the ethereal realm. This picture of the soul accords well with that of Christianity and has become a strong basis of the belief in the soul and the self, as well as the individual person, in Western thought (See also Martin and Barresi 2006, p. 4).

One might notice that I am using the words 'soul' and 'self' almost interchangeably. There are of course certain differences between the two: The soul is what inheres inside our bodies giving life and breath to the latter and makes our bodies the kind of person we are; the self, on the other hand, is strictly speaking what we refer to when we use the first-person pronoun. So in most cases the two words refer to one and the same entity. An exception is a case where a robot is capable of talking and understanding; it certainly has to use the word 'I' but in its case there is no soul because the robot is an artifact and at least it is believed that material artifacts such as a robot does not have a soul (except when there is a theory such that a soul emerges in an artifact such as a robot when it is capable of language and rational thought, but in that case the soul seems to be rather redundant). This might point to a conclusion that the soul is a subset of the self—whenever there is a soul, there is a self, but not the other way round. However, some might object that animals and plants could indeed have souls, albeit a lower kind (the Greeks themselves endorse such a theory), and if that is the case, then animals and plants do not have a sense of self as they obviously do not use the first-person pronoun. So it is possible for there to be a soul without a self after all. Here we have to be careful about our terminology. There are some research works showing that certain high level animals such as chimpanzees and gorillas do have some sense of the self. They recognize themselves in a mirror and can point and try to erase a red spot on their face that someone has earlier put there. However, we have to distinguish between the kind of sense of self which is mediated through language and linguistic understanding and the kind, as shown in the chimpanzee experiment, that is not. In order to get clearer about our purpose in this book, we will focus on the kind of self that is peculiarly human, the kind that is mediated by fully developed human language. This does not mean that gorillas and chimpanzees are not capable of communicating among themselves or even to us, but since their means of communication is limited and since the scope of this book is circumscribed on the problem of the sense of the self in the online world, then we have to limit ourselves only to the human sense of self. One might look at the difference here as one between the linguistic self-understanding or self-consciousness on the one hand and non-linguistic or pre-linguistic awareness as shown in chimpanzees and gorillas (and perhaps other high level mammals such as elephants and dolphins) in the other.

But back to our story. A sharply different picture of the self emerges with Aristotle. Although he keeps largely intact the idea that the self, qua the subject who perceives and thinks, is unitary and independent, he maintains that the self is material in the sense that it is the body in its capacity of functioning as a self or a soul that is the crux of the matter. Hence there is no independent, immaterial soul for Aristotle. According to Joseph Owens, awareness is always awareness of external objects, and self-awareness only occurs occasionally (Owens 1988, p. 707). One is always aware of oneself through one's awareness of external things, and this means that one always has to represent one's own self as spatially extended and having duration in time. In the cognition of one's self, furthermore, one is also aware of oneself as a unitary subject, and Aristotle says that this is the man himself (*anthrōpos*), and not the mind, the self, or the soul (Owens 1988, p. 707). To illustrate how this is possible we have to look closer at Aristotle's theory of cognition. It is quite well known that for Aristotle cognition takes place when the cognizing mind becomes one and the same as the object of cognition. When one perceives something, say a tree, the mind that does the perceiving somehow becomes the very tree that is being perceived. This is in stark contrast with the representational picture of cognition with which we moderns are familiar. For Aristotle, the form of the object perceived becomes the form of the perceiving mind, so there is identification between the two. In this sense, then, the mind is totally transparent, since it becomes so many things that are being perceived and cognized throughout the course of the day. According to Aristotle in *De Anima*, "Let us assert once more that in a sense the soul is all existing things. What exists is either sensible or intelligible and in a sense knowledge is the knowable and sensation is the sensible" (Aristotle, *De Anima* 3.8.432a4-9, quoted in Owens 1988, p. 709). So in a sense the self is exactly speaking "all existing things." There is absolute identification of the knower and the

known, and much of the epistemological conundrums that have bewildered epistemologists since Descartes would be gone forever if they had adopted Aristotle's
picture here instead of the representational one. But of course Aristotle has to
explain what is meant by saying that the knower becomes one and the same as the
known. The knower is the known in potentiality. When the actual act of cognition
and perception takes place, the knower's identity with the known is then actualized.
Hence there can be no way to know what the self or the soul is like without its actuality, that is without its engagement with the external world. The two, internal and
external worlds, are so interlinked in Aristotle that separation becomes almost
inconceivable. We could say that for Aristotle the Self's actuality consists in its
action, its engagement with the world: "Hence the mind, too, can have no characteristic except its capacity to receive. That part of the soul, then, which we call mind
… has no actual existence until it thinks," he says in *De Anima* 3.4.429a21. The
hand is not actually a hand until it does what a hand is supposed to do (grasping,
holding, and so on); similarly, the horse is not actually a horse until it does the same
(running, galloping, etc.). Likewise, the mind is not what it is until it thinks, or does
what a mind is supposed to do. The self is actualized when it engages with the world
or otherwise does things that a self is supposed to do, such as, presumably, thinks,
perceives, feels, dreams, plans, hopes, and so on.

Hence, on the one hand Aristotle's self is something material. There is no self-
existing soul sitting there to be discovered and described. When the self is there
only when it does something, doing some kind of activity, and certainly there has to
be something, some material thing that does all these activities. On the other hand,
it is not entirely material either because when the material thing lays inert doing
nothing, as perhaps the dead body is doing nothing, then one cannot say that there
is the mind there. Aristotle's view that the mind or the self is neither material nor
immaterial has some repercussions in our modern conception of the self in the
online world. This would be tantamount to having a self conceptualized in such a
way that it owes its ontological status to its activities in such a world. This conception might not be too farfetched, as we shall see in the next chapters when we discuss how some of these conceptions of the self shed light on the situation of the self
in the online world. As for Aristotle, when I see a tree and has an understanding that
there's a tree before me, what happens is that my mind somehow becomes the tree
I am perceiving. The form of the tree "informs" my mind so that both share the same
form. In this case there is no question, as we shall see in Descartes and in the modern representational theories of perception, of how to reconcile between the subject
and the object, as there is no such separation in Aristotle. We cannot single out some
inherently and independently existing entity and call it a mind or a soul because the
latter always exists dependently on the functioning, living body. The human soul is
what the human body does; in addition, the body can be regarded as a concrete
expression of what a soul can do, the latter's tool, so to speak. According to Shields,
the soul and the body are dependent on each other—the soul depends on the body
because, as we have seen, the soul is just what the body does; on the other hand, the
body does depend on the soul because the former is "well-suited to be an organ of
the soul" (Shields 2009, p. 300). That is, the body expresses what the soul is capable

of doing in the concrete, tangible world. The human being, a body-soul complex, can walk, think, imagine, discuss, etc. because the soul has a right tool for these jobs and the body itself would be mere inert matter without the soul. This means that Aristotle is uncomfortable with the doctrine of transmigration of the soul. When the soul depends on the body for its functioning, it is incoherent to conceive of the soul as capable of existing independently with the material tool with would make its expressions possible.

2.3 The Self in Modern Western Philosophy

The Western conception of self faced a momentous shift with the thought of Descartes. What happened was that the shift was from the idea that the source of normative rightness was outside of the individual self to one where it is the individual alone who is responsible for being right or wrong. In Aristotle's picture, one gets it right perceptually when the form of the object perceived becomes the form of the perceiving mind; in a way the mind and the object become one and the same. However, what emerged during the formation of modern scientific thought and indeed the whole of modernity was that the source of normativity was located inside the subjectivity, the thinking mind, of the individual self. Instead of the identity of things perceived and the perceiver, the moderns introduced the representational theory where the role of the mind is to represent external reality. Thus a gap was introduced which is uncollapsable and ultimately unbridgeable—the gap between the subject and the object. In Aristotle as in Plato, the role of knowledge is for the thinking mind to become one with the external reality. In fact to say that it is *external* reality is quite incorrect because without the representational theory there is actually nothing to be external to. For Plato, true knowledge is achieved when the individual mind contemplates the reality of the World of Forms and gets connected to it, thereby becoming a member of that World itself. The individual self achieves its true identity with this merging with the reality that has been the groundwork of everything all along. For Aristotle, it is the identity of the perceiver and the perceived as discussed earlier. In both cases, what is ultimately real, the bedrock against which everything else is judged, is this ultimate reality which is located outside of the individual mind. With Descartes, however, things are positioned backwards. Instead of deriving certainty and what can be taken to be the bedrock reality from outside of the mind, it is the mind itself, the subjective, individual mind of a human person, that functions as the sole source of normative rightness in both ethics and epistemology.

This shift to the subjective individual occurs hand in hand with the rejection of the teleological model of nature. The ancients viewed the world as imbued with mental qualities—in fact this is the way ordinary people view and talk about nature all along. We talk about water upstream "wanting" to flow down, or that it is natural tendency of fire to rise up, which implies that the "natural home" of fire is up and above, and so on. To the modern mind these ways of talking can only be metaphorical, but for the ancients they regarded these talks as almost literal, as if fire actually has its natural home in the air above. The emergence of modern scientific belief necessitates that these kinds of teleological talks and mindset be done away with, and a result is that mind and matter become totally separated. It no longer makes any sense to talk about water "wanting" to flow downhill; every movement is mechanistic. Instead of Aristotle's four causes—formal, material, efficient and final—only the efficient cause remains. Nature ceases to be imbued with any inkling of the mind or mental properties. In this picture, then, it is no longer possible to position the source of normative rightness in nature, the only place where it can be located is inside the subjective, thinking and feeling individual self. This is a clear source of the modern mind, one that is followed by the idea that nature is there, in Heidegger's term, "standing reserve" to be used at any time by humans (Heidegger 1977, pp. 307–342). As a standing reserve, nature is completely under the control of humans. Such control and prediction is only possible through the modern scientific mindset, which is predicated, among other things, on this idea that mind and matter are to be strictly separated and that the source of epistemological and ethical rightness is located within the individual self.

But exactly what does it mean to say that the source of normativity is located within the individual? Descartes' famous words "Cogito, ergo sum" provide an answer. One can only be certain that one's mental episodes—one's rambling thoughts and musings, etc.—represent true knowledge when they are guaranteed by God, which is only possible when the thoughts are "clear and distinct." We cannot be but are certain that these thoughts are ours and that the thought that "I think, I exist" is one where we cannot be deceived by even the most powerful evil demon. Instead of locating the source of rightness in the metaphysical reality such as the Realm of Forms or the like, it is the Cogito and only the Cogito, that provides us with such certainty. Surely Descartes relies on God soon afterwards to prevent us from despairing over never being able to jump from the Cogito to the certainty of everyday, mundane knowledge, but that comes later. The important moment is the Cogito, where we, as individual selves existing singly apart from one another in our own private sphere, achieve the source of epistemological certainty which is unshakable. God comes later as the guarantor that our episodes do represent real things outside, but God is not the source of our certainty of our own subjective thoughts anymore. In the words of Charles Taylor, "What has happened is rather that God's existence has become a stage in *my* progress towards science through the methodical ordering of evident insight. God's existence is a theorem in my system of perfect science. The centre of gravity has decisively shifted" (Taylor 1989, p. 157).

With Descartes we see the emergence of the modern individual self, one that plays an important role in liberal thought and indeed a foundation of contemporary discussions in social and political philosophy, as well as in legal theory and indeed in much of what is being deliberated in the West on information ethics. In any case, the total separation of mind and matter in Descartes results in the mind becoming wholly immaterial, and matter becomes wholly extensive and inert. This is not the same as the immaterial soul in Plato and in the ancients because the immaterial soul, even though immaterial, nonetheless has its own entity and existence in its own right; in other words, the ancient immaterial soul can be conceived as an object. But the Cartesian soul cannot at all be conceived as an object. It is a pure subject through and through, and this view is peculiarly modern. The existence of the Cartesian soul is only through the first-person perspective; it exists only as a subject, the thinker herself, and cannot be conceived of as any kind of an entity or an object through some perspective of another. In this case the gap between the subject and object is clear, and it is clearly unbridgeable. It is not surprising, then, to see the Cartesian self play an important role in conceptions of online selves which are detached from their physical bodies. Ray Kurzweil once proposed that we should download the content of our minds or brains on to a server in order to achieve immortality (never mind the possibility that the server may go down or the content accidentally deleted because of power failure, etc.) (Kurzweil 2005). This is clearly derived from the Cartesian vision. What there is to the mind and to the self is nothing more or less than its informational content. This is a powerful vision, and one that is very influential in current thinking about online selves. We shall see in the next chapters how this plays out in more detail, but for our purpose here let us go to another philosopher who is famous for maintaining that the mind and the body are in fact one and the same and hence the two cannot be separated at all.

Spinoza is famous as a radical monist. He maintains that everything is essentially one, the one substance which is God or Nature depending on your choice of words. He repudiates the Cartesian dictum of the radical distinction and maintains that both the mind and the body are "affects" of one and the same God or Substance, which is necessarily one. For Spinoza the mind is "the idea of the body." The idea is that the mind and the body exist in parallel to each other. The mind that belongs to body assumes the shape and form of that body. For example, my own mind reflects my own body; thus it resembles my own body, looking like me and having a voice like me and so on. In Descartes' theory this is of course impossible because since the mind is immaterial it cannot assume any look and shape. But if it can't do so then it becomes utterly mysterious how my mind differs from yours, for example. According to Descartes the mind is totally immaterial and non-extensive, so there are no distinguishing marks that would make one mind different from another. The mind in Descartes' theory, then, becomes totally indistinguishable one from another. Thus it appears that Descartes' theory is more mysterious than Spinoza's. Spinoza's view that the mind is the idea of the body could be taken to mean that the mind is a kind of mental correlate, so to speak, with the body. God is the one substance, the only one in existence; ideas are modes of God, belonging to the mental attribute which expresses God's essence. Since God is infinite, all possible ideas are in God. This includes the ideas that are directly conceived by God and all other ideas that are derived logically from the original ideas. This means that all and any possible ideas are in God and owe their being ultimately from God. Furthermore, Spinoza maintains that there can be no interaction between the mind and the body. Even though they are both derived from the essence of one and the same God, they belong to different attributes of God and thus cannot cause each other to act. Spinoza says in the *Ethics*, Book II, Proposition VI: "The modes of any given attribute are caused by God, in so far as he is considered through the attribute of which they are modes, and not in so far as he is considered through any other attribute" (Spinoza 1985. All quotations from Spinoza are from this volume). God has an infinite number of attributes, of which we humans can conceive only two, namely body and mind. Modes or individual things belonging to one attribute, such as individual bodies, which belong to the attribute body, and individual ideas, which belong to the attribute mind, cannot interact with modes belonging to the other attribute. In other words, bodies cannot be affected by the mind, and vice versa. However, Spinoza also maintains that bodies and minds are essentially one and the same, because all belong in God, who is absolutely one and the same. So in the end both body and mind are only expressions of one and the same entity; they only appear different only in their capacity as modes of either the physical and mental attribute. According to Spinoza, "I wish to recall to mind what has been pointed out above—namely, that whatsoever can be perceived by the infinite intellect as constituting the essence of substance, belongs altogether only to one substance; consequently, substance thinking and substance extended are one and the same substance, comprehended now through one attribute, now through another" (*Ethics*, Book II, Prop. VII Note). Furthermore, in the same Proposition, Spinoza states that "The order and connection of ideas is the same as the order and connection of things" (*Ethics*, Book II, Prop. VII). There is a parallel between the mental and the physical, such that the one is an exact copy of the other. However, the parallel is only a result of our own limited way of conceiving the two attributes, for ultimately they are one and the same.

In Proposition XI of Book II, Spinoza writes: "The first element, which constitutes the actual being of the human mind, is the idea of some particular thing actually existing" (*Ethics*, Book II, Prop. XI). The actual being of the human mind is constituted through the idea of some particular thing which is actually existing. This is possible because the human body is complex enough to accommodate the kind of mind that is capable of complex thinking that is typical of a human being (Book II, Prop. XIII). The human mind would have been nothing if not for the existence of a particular thing that is being thought of by the mind. Thus the two, mind and the thing conceived, are always dependent on each other. On the one hand, Proposition XI says quite clearly that the mind is constituted through thinking of some particular thing; it is the result of the act of thinking which is directed at a particular thing that makes up the human mind, so to speak. On the other hand, without the mind it is impossible for there to be any particular thing either. This is because everything is already being conceived of by the infinite intellect of God, so the parallel between mind and body is pre-existing; hence it is impossible for there to be something bodily without anything mental as its correlate. Spinoza says, "… that whatsoever can be perceived by the infinite intellect as constituting the essence of substance, belongs altogether only to one substance: consequently, substance thinking and substance extended are one and the same substance, comprehended now through one attribute, now through the other" (Book II, Prop. VII Note). That is, each particular thing is an idea and each idea is a particular thing, the difference being only that it is conceived through one attribute or another. In a sense each particular, material object is simultaneously an idea, at least in God's perspective; however, we also have ideas which are our own thoughts of particular things or subject matter. We are aware of our ideas as soon as we think of something. This is the human idea, and it is distinguished from material objects as ideas of God by having the form of thought which is known as such directly to us. In other words, for Spinoza the mind is a parallel of the body; the structure of one always reflects that of the other. Furthermore, the two are only parallels of the other only when conceived of through different attributes. In reality they are one and the same. A material object such as a rock is at the same time an idea. There has to be an idea corresponding to each and every material object, even if that object is not conceived of or perceived by any human being, for everything must be in God. However, an idea that does have a physical correlate is known as 'confused' or 'inadequate' idea. For Spinoza no idea is actually confused or inadequate because God conceives of everything; hence it appears confused and inadequate only in relation to a finite human being (Book II, Prop. XXXVI).

What all this means is that the mind for Spinoza is one and the same with the whole of reality. This is very difficult to imagine to the modern sensibility, but this is exactly what Spinoza means. God is the whole of reality, and it (for this is about the only way to refer to Spinoza's impersonal God) can be conceived of by the human mind only in two ways, namely as mind or matter. The human being is a complex consisting of both body and mind bundled together; in fact to say that the two are bundled together is not all together accurate, for they are one and the same. The human mind for Spinoza then is nothing but the human body. The body is the mind manifesting itself in the material world, so to speak, or to put it in another way the human body is the human mind itself. The reason why we seem to have a lot of difficulty understanding Spinoza's view here is perhaps that we are so accustomed to the Cartesian radical distinction that we may be at a loss to find out what it means to say that the mind is in fact the body. But viewed from Spinoza's vantage point of the "point of view of eternity" (*sub specie aeternitatis*) this is exactly what he means.

2.3.1 Early Modern Philosophy and the Online Self

Spinoza's view that the mind and the body are one and the same has very strong implications for our goal of understanding the online self. The implication is that the online self, then, is both body and mind at the same time. Suppose Spinoza's view is tenable, which I believe it is and will argue for this point in the next chapters, the mind does not have to be limited to the extent of the normal body. As the body can be extended through the use of prosthetics or other devices, so can the mind. Andy Clark and David Chalmers (1998) argue for the thesis of the extended mind, one which I find to be very congenial to the Spinozistic viewpoint here. In this sense the online self can be thought of as an extension of the human body, the body of the person who creates and operates that online self, then as the body and the mind are one and the same, the mind is there in the online self too. Or to put it perhaps more accurately the online self itself is the online mind. This is different from the traditional Cartesian attitude which may look at the online self merely as an expression of the immaterial *res cogitans*. In fact the online self is also physical; what is physical can be found in the usual way of people seeing and interacting with the online self through a variety of sensory means, chief among which, of course, is the visual, but there is no principle that would prevent in principle extensions of the interaction with the online self through other sensory means. So the online self is both physical and mental at the same time, just like the usual self or the usual person in the offline world. We shall certainly discuss this topic, which is central to the book, the later chapters.

While Descartes and perhaps Spinoza may be regarded as precursors of the modern liberal self, this kind of self actually comes into being with the philosophy of John Locke. Chief among Locke's ideas on the self is that it is composed of ideas and, famously, that personal identity is constituted through continuity of memory and consciousness. Ideas for Locke are individual and mental representations of individual things in the outside world. When one perceives a tree in front of one, for example, one directly perceive a representation of the tree that one is aware of in one's "mind's eyes" so to speak, and then only by inference does one come to perceive that there is a tree in front of oneself. In terms of personal identity, Locke holds the famous view that it is continuity of consciousness that is constitutive of personal identity. It is one's own memories, accessible through one's own subjective recall, that enables one to collect these memory episodes together to form a sense of one entity that is one's own self. In other words, for Locke the crucial factor in binding different mental episodes together is the first-person, subjective awareness that it is oneself, that it is one's own person that exists through time as recalled through one's use of memory. This is a key to another famous view of Locke's that the self is autonomous and atomic. The ability of the person to reflect in the first-person, creating a field of subjective awareness in which one becomes aware of oneself as separated from all others is the key toward the modern idea of the atomic individual which plays an important role in the liberal theory. According to Locke, the self is "that conscious thinking thing, (whatever substance, made up of whether spiritual, or material, simple, or compounded, it matters not) which is sensible, or conscious of pleasure and pain, capable of happiness or misery, and so is concerned for itself, as far as that consciousness extends" (Locke 1997, p. 307). The key point here is that the self is "concerned for itself, as far as consciousness extends." This shows that it is the self's reflective awareness and being conscious of oneself as a self, that is the issue. In Descartes, the cogito self is aware of its own thinking, but it does not have to be aware of the fact that it is an individual self in contrast to all others. After all, the Evil Demon could have created the world such that there is only one individual person in the whole world who is contemplating the cogito moment; there would not be any inconsistency in such a world that the individual 'I' there be only one thinking person. However, for Locke, the fact that one has to rely on one's memory and subjective continuity implies that one is aware of oneself as being extended in time. And as it is obvious that in one's narrative one has to engage with others in one way or another, one gains a sense of separateness, of being one's own self in contrast with others, which implies that others do exist. This seems to be a key difference in Locke and Descartes. Narratives have to be populated with characters; it just does not make sense to cook up a narrative with only one character who does not interact with anyone at all because he or she is the only one there is. Locke's world is a commonsensical one where one interacts with one's peer and so on; even if there were an Evil Demon Locke's theory still goes through because he does not aim at complete certainty and refutation of the Evil Demon, but only that one creates a sense of oneself by constructing a unique narrative. That the narrative in question might be hallucinatory in the all-encompassing way as required by Descartes would seem to be irrelevant.

2.4 Kant

Locke's picture of the individual and atomic self is taken up by Kant and develops into a full blown autonomous self. In Kant the self develops into a kind of self-legislating and self-determining subjective entity that is a member of the "Kingdom of Ends." Epistemologically speaking, Kant cannot argue that the self is situated on a firm basis; that is, Kant admits, in the *Critique of Pure Reason,* that one cannot find any scientific or empirical evidence to prove that the self actually exists as an entity that can be ascertained through the third-person perspective. The self can only be known through its role in the form of every judgment. In Kant's term, "The 'I think' must be able to accompany all of my intuitions." That is, in every act of thinking, there has to be the subject of such though in such a way that the thought be a coherent and meaningful one. The 'I' in the 'I think' that has to accompany all of a person's thinking episodes, then, cannot be regarded as a substance on its own because it has to remain always a subject. To regard it as an object, which is necessary for it to be a substance, would imply that one steps into the realm of the unknowable which is forbidden in Kant's own system from the beginning. In other words, the 'I' in the 'I think,' namely the logical and formal subject of each and every thought, cannot be regarded as an object of any kind, which means that it cannot be brought into the realm of knowable individual objects, the realm that the *Critique of Pure Reason* aims to investigate. To investigate the subject of the 'I think' as if it were an object would for Kant be tantamount to encroaching upon the realm of the unknowable, since to know anything presupposes that the thing known has to defined through the system of categories that locate the thing in the knowable world. Hence, the metaphysical self, the self which is the pure source of subjectivity, cannot be known empirically or scientifically; the self in this sense, which Kant terms the Transcendental Apperception, exists only in the formal sense, as literally the 'I think' that accompanies all of someone's intuitions. To try to bring the Transcendental Apperception to the realm of the knowable would also involve a vicious infinite regress, because for a supposed 'I think' that is brought down and made into an object, there has to be another 'I think' that looks at and thinks about it, and when that is brought down, there has also to be another 'I think,' and so on and on.

The problem of how to combine the various episodes so that they belong to one overarching self is well known. Kant posited the "Transcendental Unity of Apperception" as a means by which these episodes are combined so that they belong to one and the same subject, which would make cognition (or in his words "judgment" and "understanding") possible (Kant 1929). However, a problem with the Transcendental Unity of Apperception (TUA) is that it is a purely formal concept, and does not contain any particular information that pertains to any particular individual. According to Kant, "it must be possible for the 'I think' to accompany all my representations" (Kant 1929, B131–132), meaning that it must be possible for me to be conscious of all my mental episodes; otherwise it would not be possible for me to be justified in asserting that these episodes are mine. But what is very interesting here is that Kant is not arguing here that there must be an objective self, the "I" which "thinks." Kant is putting forward a transcendental argument here. A transcendental argument is one that accounts for a condition of possibility of a certain phenomenon, its point being that, for the phenomenon to be an objective one at all, or for it to be even possible, certain conditions must already obtain. The transcendental argument does not show tout court that the phenomenon exists objectively; that would run against the spirit of the critical philosophy. For example, in arguing about causation in the Second Analogy, Kant's point against Hume's devastating attack on causation is that there are conditions of possibility of causation, such as if an event A were to be the cause of another event B, certain conditions need to obtain, such as that both A and B need to be able to be subsumed under the pure concept of understanding of logical relation. Kant does not rebut Hume directly; he does not argue that Hume's argument is directly false. What he does is that he argues that, if we are to be able to maintain objective knowledge, we need to posit the concept of cause and effect. Hence the concept does apply objectively to phenomena. In the same vein, Kant argues that it must be possible for the 'I think' to accompany all my representations. He does not argue categorically that the "I" has to exist; what he says is that the "I" needs to exist as a condition of possibility for relating disparate representations into a coherent whole, which in turn is necessary for there to be objective knowledge. What the "I" is doing here is nothing more than a place holder, a formal factor that serves to unify various representations together so as they belong to a coherent self.

This is of course not a place to examine Kant's philosophy in any detail. But if the "I" of the Transcendental Deduction here functions as a purely formal unifier, then this "I" would be devoid of all and any characteristic that would qualify it to be the "I" of any particular person whatsoever. All it can do is to perform this purely formal function, which must be the same for everybody. In short, the "I" here functions as the Transcendental Unity of Apperception (TUA). Thus, my TUA is exactly the same as your TUA, since both function in the same way and cannot contain anything unique to either me or you. Anything unique would be empirical and cannot be part of the TUA. If this is the case, then Kant's TUA is too general and cannot perform the work expected of the individual self. In other words, one cannot rely on Kant's transcendental argument about the "I think" here and use it to argue that the self does exist as an objectively existing being.

Since any attempt at finding the overall unifier of the mental episodes would fall under the empirical side of things (because once a candidate for the unifier is identified, it then falls under the category of a mental episode which is being thought of, which then requires another subject to think about it, and so on), or under the purely formal schema such as Kant's, which is empty. An upshot, then, is that any attempt to bind up the episodes is always provisional and cannot escape from being itself yet another mental episode. When one attempts to bind up one's own episodes, one is then conscious of yet another episode whose content is about the binding, but then that becomes another mental episode in need of further binding. Consequently, the offline self is a construct in the sense that it is not there objectively or ontologically. It is something "made up" in order to facilitate daily living of any human being. For example, it would be much easier for me to refer to you, using your proper name, if you stay relatively stable throughout some period of time, even though analysis shows that there is ultimately speaking to real "you" in the ontological sense. What I and others take to be "you" is a social construct not dissimilar from Searle's example of a bank note whose value is also a social construct (Searle 1997). In other words, the value of the bank note does not reside ontologically in the material itself, but sociologically through agreement among members of society that this particular type of a bank note has such and such monetary value. In the same vein, when I refer to you, calling you by name for example, I am abiding by certain social conventions that recognize that, relatively speaking, there is a certain person behind the persona that I am now perceiving.

2.4.1 Kant and the Online Self

But if this is the case, then it is also similar for the online self. We can look at the online self as a persona that the individual makes up as a front to present himself or herself to the world, and sometimes the person may intend it in such a way that the persona assumes identity of its own, without being able to refer back to the real person behind. The online self is also made up of physical and "mental" episodes. The physical episodes are easy enough to understand—bits of electron working together to present images, sounds, and texts on screen. But the mental episodes are also there, as we can gauge what the persona is thinking or feeling through her use of language and other symbols (such as emoticons) through the Internet. These episodes also need to be connected together in order for us to form a more or less coherent picture of a self working behind. Here one also finds an analog of the Transcendental Unity of Apperception in the online world too. Just as in the offline world, the analog of the TUA in the online world functions to bind the different episodes of postings and comments together so that they belong to one person. It thus functions more as a regulative agent working as a condition of possibility of there being a coherent self behind the various representations constituted through images and texts that are posted online. So the analog is something like this: "It must be possible for the 'I think' to accompany all my postings of links, images, videos, comments, etc. on the social networking site; otherwise no coherent self does not emerge which is necessary for there to be social networking at all." But since the offline self is ultimately speaking a construct, so is the online one.

An interesting upshot of Kant's view on the TUA discussed here is that the TUA seems to be successful in identifying the self or by implication the person. It seems to succeed in binding up various representations together so that they belong to one and the same self. However, it does not seem to succeed when it comes to securing uniqueness of a particular self or of a particular person. What the TUA does is that it gives me, for example, a means by which I can be certain that I am a coherent self and that all the representations that are flitting across my brain or my cognitive field are indeed mine. However, what the TUA does not do is to identify me as someone who is distinctly different from another person. Since the TUA is a purely formal apparatus, it cannot do this job, because what makes me a unique person, such as someone who is teaching philosophy and who is interested in many subjects and so on, cannot be contained in the TUA function that I have. Identity is not the same as uniqueness. There can be several things each of which are of course identical to itself, but without uniqueness all these things are just a bunch of entities sharing all properties in common (Leibniz's Law notwithstanding) but having no unique iden-tity of its own (i.e., the characteristic of being itself alone and none other). This is not something that can be accomplished by the TUA. Hence, in order to account for my uniqueness, external factors need to be considered too. For example, in order for me to be certain that I am unique, I usually refer to the set of characteristics that only belong to me and are shared by no one else. Since the TUA functions in exactly the same way in all the selves and all the persons, it cannot specify uniqueness. And since Kant's view on the TUA appears to be the best shot given by the internalist theory of specifying identity, we also need external factors too, at least when it comes to uniqueness.

In the chapter of Paralogisms of Pure Reason, in which Kant discusses problem in the theory of mind and self, he shows that there are always two layers of the self—the apperceptive 'I think' that accompanies all thinking episodes discussed earlier, and the empirical self which can be subject to empirical or psychological investigation. The former cannot be individuated in any way, since individuation presupposes the pure concept of individuality and plurality, which is not available for unknowable entity; the latter, on the other hand, is empirical in the sense that it can be described in ordinary language. According to Kant, "[i]n all judgments I am the *determining* subject of that relation which constitutes the judgment. That the 'I', the 'I' that thinks, can be regarded always as *subject*, and as something which does not belong to thought as a mere predicate, must be granted. It is an apodictic and indeed *identical* proposition; but it does not mean that I, as *object*, am for myself a *self-subsistent* being or *substance*. The latter statement goes very far beyond the former, and demands for its proof data which are not to be met with in thought, and perhaps (in so far as I have regard to the thinking self merely as such) are more than I shall ever find in it" (Kant 1929, B407). The statement "The 'I think' must be able to accompany all of my beliefs and judgments" (Kant 1929, B131–132) is for Kant analytic and necessarily true because the very act of thinking, having a relation to any proposition at all, requires that there be a subject to relate to. The relation has to be between the subject and its object, namely the proposition. On the other hand, when Kant says "I, as *object*, am for myself a *self-subsistent* being or *substance*" what he means is that when the self is examined empirically, such as when I look at myself and investigate whether some of my current beliefs are true or not, the self here is entertaining a synthetic proposition, one which obviously cannot obtain its content through analysis and thought alone. When I examine myself whether some of my beliefs are true or not, the situation is peculiar to my case alone. It is my own life history and experiences that partially explain why some of my beliefs might be false or at least merit some serious investigation; the fact that it is *my* own life history shows that the empirical self in question here belongs to the form of synthetic proposition and as such is subject to the knowable realm of the *Critique of Pure Reason*. The two layers of the self here is meant by Kant to show that any attempt to give an account of the metaphysical self as if it were a knowable object will necessarily conflate the two layers here and thus will always be destined to fail.

Here Kant and Locke share a conception of the self which is fully individual and atomic. The self is the purely subjective entity, the subject of the 'I think' which necessarily accompanies all thoughts of the subject. In Locke's term this means that the self is the sole subject of his or her own subjectivity, the thread that ties the various memory episodes together to form a coherent narrative that is constitutive of one's own sense of self. For Kant, he somehow makes way for the self to be become the noumenal subject; that is, a subject that is free from the channels that are necessary for any conception and judgment of the knowable objects. It is this noumenal self, according to Kant, that is the source of freedom that is crucial for constructing a theory underlying a system of ethics and political thought. It is the self that underlies the "good will" which is the subjective awareness of oneself as the author of one's own action who acts out of an awareness of reasons as to why she acts the way she does. These reasons are then put under the rubric of maxims. One acts rightly, according to Kant's system of ethics, when one acts in such a way that any maxim that explains one's decision and resulting action follow such a maxim that is universalizable into general law. This is first of Kant's famous "categorical imperatives," a set of rules that lay out clearly what it is to act ethically. All this is only possible because the self is absolutely free and rational; without freedom and rationality, no *ethical* action can take place. Another of Kant's categorical imperatives says that one should act in such a way that one becomes a "fully legislating member" of the "Kingdom of Ends" (Kant 2012; see also Korsgaard 1996). One is only a fully legislating member when one is fully autonomous; when one, that is, is totally free and rational and acts out of one's full understanding of any and all reasons that are relevant to the decision and action. Kant envisions a "Kingdom of Ends," a kind of imaginative assembly where the members are fully free and autonomous in this sense. Any agreement on rules that results from a deliberation of the assembly then would have a normative force that is binding on each of the member. This is the basis of Kant's political theory and of democratic liberalism in general. Rawls' view of a theory of justice based on an assembly of individuals each of whom operate under the "veil of ignorance" is derived directly from Kant's view here (Rawls

1971). The Kingdom of Ends are such because the members of inhabit it are fully autonomous, rational and individual (in other words, atomic). As such they are ends in themselves because they act out of the interests of rationality and autonomy alone and not out of any heteronomous motives. The fact that the injunction to act as if one is a legislating member of the Kingdom of Ends is itself a *categorical* imperative makes it necessary that any action arising out of following this rule would presuppose that the self who does the decision and action always acts as an end in herself rather than a means (Kant 2012). The concepts are all interlocking with one another: In order to be autonomous, one has to be fully free, and freedom and autonomy are the basis on which rationality is possible; furthermore, one can only be an end in oneself if one is rational and autonomous in this sense. Since freedom, autonomy, rationality and being an end are all constitutive of the good will that Kant says at the beginning of the *Groundwork* (Kant 2012) to be the only thing in the entire world that is capable of being good in itself, these concepts are all constitutive of the ethical.

The key idea for us here is the view that the self is autonomous, fully free and rational. This is only possible if the self is an atomic individual as we have seen. Without being atomic in this sense, it is very difficult to envision how such a non-atomic self could act in an autonomous manner, as a fully legislating member of the Kingdom of Ends. The idea that the self is an end in itself also points to the fact that the self, the individual human being, has dignity and deserves respect. Hence comes the view that human beings are to be accorded with a special kind of treatment and that they are endowed naturally with a set of rights, the human rights. All these are basic to the modern liberal theory. One sees, then, how hugely influential Kant has been in the whole of ethics and political theory that ensues.

Thus both for Locke and Kant the self are atomic and autonomous pure subjectivity, the subject of thought episodes that can be threaded together through memory. This pure subjectivity becomes then the source of human beings as ends in themselves and thus the idea that humans deserve special attention and treatment. In other words, pure subjectivity renders humans to be not of this world, since in Kant the transcendental apperception does not belong to the sensible world. Since freedom cannot be found anywhere in the sensible world, it is taken to be one and the same as the pure subjectivity. The Lockean and Kantian self is nothing if not pure freedom, or pure moral agency. It is autonomous because, as we have seen, the self makes law unto itself through a system of rational ethical decision making; it is also atomic because the process of making the decision ethically depends on the self alone and requires none other in the effort. This is the source of the liberal self. From there we get the familiar picture of the conglomerate of selves getting together to form a social contract which is the basis of the legitimate use of political power in a polity.

2.5   Online Self and Liberal Self

As for the online self, the liberal self has a large role to play. We will examine this in the later chapters. Apart from the familiar ethical picture in liberal theory, one that rests on the basis of the Kantian and Lockean self, an important issue that will clearly be taken up in detail later, one finds, metaphysically speaking, that the real self is nowhere to be found because it lies outside the scope of any empirical investigation (unless of course it is there as the subjective first-person source of thought for the subject). The self that is there in the online world is only a manifestation of the subjective self. What is there on screen is an outward manifestation of the internal subjective self. Here one finds a sharp distinction between what is inside and outside. The profile of someone's self as appears on social media websites such as Facebook is a persona, a projection, of that person's internal self. What is important, then, is the internal self, and the persona is merely that, a mask that presents the hidden internal self to the outside world. This view underpins Goffman's famous analysis of face-to-face interaction as theatrical performance (Goffman 1959). Since the online self is a mask, the real issue then lies with the internal self, and this of course implies that there is no real, independent analysis of the online self as a separate phenomenon. Analyzing the online self would be little more or less than analyzing someone's masks which she puts on in various occasions. The masks can then represent the wearer in one way or another, but always the real person behind remains one and the same.

2.6   Hegel

As promised in the introductory chapter, I shall show that this picture of the internal self wearing a mask presenting herself to the outside world is a misleading one because it presupposes that there is a hard and fast line separating the two (See, for example, Ess 2010). This separation is contrary to both Spinoza's and Hegel's view of the self. Thus we turn now to Hegel's view and then an argument as to why the Spinozistic and Hegelian view is superior to the Cartesian one that is based on the strict separation between the mental and the physical.

Hegel is well known for many things, and one thing he is well known for is his critique of Kant's ethics. Basically put, his critique is that Kant's ethics is formal and does not contain the kind of substance that would inform us how best to live and would correspond to the changing historical circumstances (See, for example, Geiger 2007). The central point of Kant's ethics is with the categorical imperatives—one should act in such a way that follows a set of rational, universalizable rules. Kant's ethics does not tell us in any substantive way what exactly are the acts that should be done or should not be done. This is a quintessential modern thought, one that allows for a variety of instances while controlling the form or the constraint that those various instances must take. Hegel's contribution to ethics is that he tries to bring back the ancient vision, now as a foil against Kant's abstract and modern system. Based on his elaborate and dynamic metaphysical system, Hegel situates Kant's ethics midway between simple egoistic desire and one that totally integrates the interest of an individual with that of the community. In simpler terms, Hegel views Kant's system as essentially incomplete: While Kant's view is a step above the simple egoistic demand of the kind "This is good because I want it," it is still predicated upon a separation between the individual (the rational self who is a member of the Kingdom of Ends, as we have seen) and the community, which for Hegel is the absolute precondition of the very existence and meaning of an individual in the first place. In the ultimate vision, Hegel views a system of ethics as a total merging between the individual and the community. This is not to say that the individual completely loses herself in the faceless sea of community; on the contrary, the individual, in fact all the individuals, is still there within the community, but her individuality and her existence becomes an organic part of the community such that her interests and those of the community are inseparably intertwined. To be sure, Kant's system also recognizes the role of others and the community: When one follows the first Categorical Imperative ("Act in such a way that your maxim is universalizable as a general law"), it might seem that others are required because it is their reciprocal act, their acceptance also of the Imperative, that concretely manifests that the maxim is indeed universalized. However, this is not absolutely necessary. Even if one exists alone in the world, one is still able to follow the First Imperative here if the maxim explaining one's decision and action is in principle universaliz*able,* if, that is, the maxim would have been adopted also by other rational individuals reciprocally were they actually around. But the actual fact that others are around is not necessary: If the maxim can be universalized in principle, then it is enough in Kant's system.

However, this is not enough in Hegel's system. Since the individual is ontologically constituted through her relation to other individuals in the first place, there does not seem to be the question of whether it is possible in principle for an individual to be utterly alone or not. Even if an individual were indeed alone, such as when she is a cast away on a remote island, the actual, concrete fact that she is what she is materially and personally only through her upbringing and socialization by others is constitutive of who she is. Any action she is doing has to be judged by others; it is not enough if she merely follows the Categorical Imperative. In order for her action to be able to be judged as bearing an ethical value, others have to do the judging. This requires that they have to be around. Thus when she is cast away on the island, there is a real sense in which whatever she does does not have to have an ethical value. Since for Hegel the real and the rational are ultimately one and the same, the ethical then has to be both real and rational. This means that the ethical has to be situated in the real situation where there is a community of individuals in order even to become ethical in the first place.

Another way of looking at the difference between Kant's and Hegel's conception of the self is that, as previously discussed, Kant's view of the subject is totally impersonal. The subject of ethical decision making, the 'I think' that accompanies thoughts, desires, and conscious action is devoid of any marks that single out one particular individual from another. For Kant such a mark is not possible in his system because it has to be devoid of any such marks from the beginning because he is dealing with an abstract system which forces him to entertain a conception of the self which is wholly formal. Thus for Kant there is always a distinction between the empirical and the transcendental. The former is knowable within the realm of the senses and is subject to the categories. The latter, on the contrary, is entirely outside of the knowable realm and acts as a condition of possibility of the there being the former. For Hegel, this distinction, as is the case with any distinction for that matter, needs to be overcome. His well-known system of dialectics posits a conflict between two opposing sides, resulting in the birth of a new thing which combines the identities of the former conflicting parties while at the same time contains new identities of their own. The Kantian distinction between the transcendental and the empirical selves, then, is engaged in the process of dialectics and result in a new conception of the self which is both formal, acting as the overall subject of thought, and empirical, having stuff that marks out one individual from others. This emerging, synthesized self then retains her individuality and personality while functioning as pure subject. As is the case with the formal Categorical Imperative, the idea is that, as the Imperative provides a formal condition that specifies the *condition of possibility* of ethical action, the transcendental self then provides a formal condition of possibility of an individual self. However, in this case the self is not the simple individual incapable of thinking of or for others any more, but a much more enriched one that includes other personalities or more accurately all identities and personalities including one's own individual self. Perhaps a good analogy to help understand this is that of a giant mosaic consisting of many small pieces that all together make up a huge image. The small pieces are all pictures of other things, but when they are put up in a certain manner they together make up the big picture.

2.7   Two Strands of the Self in Western Thought

We can see then that there are two main pictures of the self in Western philosophy. For Descartes, Locke and Kant, the self is characterized by its strict separateness from the world. For Descartes it is the radical distinction between mind and body and the idea that the Cogito is the starting point of all knowledge (which implies that such a starting point cannot have its place in the empirical world). For Locke it is the idea that the self is constituted through its memory and its narrative it makes of itself, an idea that Kant roughly shares through his view that an identity of the self or the person cannot be anything from the empirical world in the sense that it is received from outside. We shall see a critique of what I call the 'internalist' view of personal identity and its implications for the online self in the later chapters. My critique will share the same spirit as the views of the self in Spinoza and Hegel, in that the self is not separated from the world. We have seen that Spinoza argues that the individual self is akin to the individual object in that both are ultimately continuous with Substance or the basic reality that pervades everything. Hegel views that self also as continuous with the basic reality, which in his terms is called the *Geist* (the term is usually translated as 'Spirit' in English, but there are certain senses of *Geist* in German which is not captured in the ordinary use of 'spirit' in English). In any case, what Spinoza and Hegel share with each other is that the individual self is not separated ontologically speaking from basic reality, an idea which resonates with many strands of Eastern thinking as we shall see.

I have said in the beginning of this chapter that I will try to tell the story of the self, so far as it can be done in such a short chapter as this one, in a way that combines the two large strands of the West and the East together. I have spent some pages in the chapter on some of the major strands in the West, starting with the Greeks and ending, as of now, with Hegel. However, the story of the self in the West does not end with Hegel, and the philosophies after Hegel are very interesting and are actually definitive of our current predicament regarding the self. Hence, in order to interweave the stories we turn now to look at the story of the self as it happened in the East, and since the East is divided into two major civilizations, that of India and China. We see then that the task of telling the story of the self is tremendous. It is clear that to tell the story in each of the major strands either in India or China (not to mention of course the European story that we have just seen) would be a very major undertaking requiring many volumes of studies. Nonetheless, what I attempt to do here is only to provide an overview of the story as it happened in the East and the West in order to come up with what I think to be an adequate background for understanding the current phenomenon of the online self.

2.8   The Self in Chinese Philosophy

We begin with China first. It is well known that Chinese philosophy in general is not as interested with metaphysical issues as in Indian or Western philosophies. This is in tune with the general tone of Chinese philosophy that is more concerned with the problem of day-to-day living and of the immediate here and now rather than with searching from more ulterior reality. Henry Rosemont, one of the most prominent scholars on Chinese philosophy, has the following to say:

> …the Chinese philosophical terms focus attention on qualities of human beings, as a natural species, and on the kinds of persons who exemplify (or do not exemplify) these qualities to a high degree. Where we would speak of choice, they speak of will, resolve; where we invoke abstract principles, they invoke concrete human relations, and attitudes towards those relations. Moreover, if the early Confucian writings are to be interpreted consistently, they must be read as insisting on the *altogether* social nature of human life, for the qualities or persons, the kinds of person they are, and they knowledge and attitudes they have are not exhibited in actions, but only in *inter*actions, human interactions (Rosemont 1991, p. 89).

What is important in Rosemont's quote here is that the qualities of a human being are defined not through abstract properties as in Greek and Western philosophy, but through interactions among humans. In this manner it does not seem to make much sense in the context of Chinese philosophy to say anything of a human being in general, because such a general human being does not actually relate to other humans since he is general and represents every human to begin with. On the contrary, the qualities of a *particular* human being are very much in question. This particular human being, then, has to relate to others in a number of ways, and these relations with other human beings in an actual social context define who he is and constitute any properties that he or she has. Taken to the extreme, this means that in Chinese philosophy there is no question of the metaphysical nature of a human being, no question of the kind "What is it to be a human being?" as in Aristotle. Instead there is the day-to-day, more mundane question of who this particular human being is—what village does he come from?; what role does he have in society?; and so on. It is this general disregard for metaphysical questions that define the general attitude of Chinese philosophy. This does not mean, however, that there is no metaphysics in Chinese philosophy at all, but it shows the general tendency of the philosophy which is geared more toward the mundane and the concrete rather than the supramundane and metaphysical.

To this extent what I am is defined exclusively by the roles I play in the society. I, for example, am a son to my parents, a husband to my wife, a father to my son, a teacher to my students, a colleague to my co-workers, and so on. The difference between Confucianism and much of Western philosophy is that there is no conception of an isolated self existing above and beyond the totality of relations that exist between myself and others surrounding me. In other words, there is no subsisting 'me' that exists apart from my roles as son, father, husband, etc., my individual soul, so to speak, that essentially constitutes who I am. I am the totality of the roles that I play (Rosemont 1991, p. 90). The notion of the self is thus much more fluid in Confucianism than in much of Western philosophy. (I said much of Western philosophy because there are strands of the latter which are more or less similar in the Confucian conception in that the self is relational—we will certainly discuss this topic later.) What is clear in the Confucian conception is that I do not choose the kind of person I am; since I am nothing more or less than the sum total of my relations, there is an important sense in which these relations are somehow imposed on me because I cannot choose who my parents are, who my son is, and so on. I always find myself situated in a local context, something that happens as a matter of course and that I have to take account of. It is merely a fact that I am born this way, and the more important aspect for the Confucian is that I have to honor this position through performing certain rituals *(li)* which reflect the way things are for me and for all others, in order that things go smoothly and harmoniously in their way. I only find out who I myself actually am only after these relations have become clearer. Only when I have a son, which entails having a wife and a set of social relations with my in-laws, and a position in society such as being a professor, and so on, does my identity as a distinctive self emerge. Who I am is in constant negotiation and relation with these social relations.

In the online situation, one can certainly analyze the self there through the Confucian perspective (Charles Ess has written quite a lot on the topic of intercultural information ethics; see Ess 2005, 2006, 2007, 2010). What happens is that one looks at the self there as the sum total of one's own social relations. A profile on

Facebook, for example, is nothing existing independently in itself, but is constituted through his or her being friends to everybody who is connected with her in the online world. Here, perhaps, lies a deficiency in the Confucian conception, for the Confucian views relations as various and perhaps hierarchical. In the Confucian world one just is not a friend to everybody; one has to be a son, or an employee, or a student, or a father, and so on and on. But what appears to be the norm in the world of Facebook and the like is that one is a *friend* to everybody in the universe. In such a case, then, everybody on Facebook would tend to be one and the same as everybody else because everyone has same kind of relations, that of a friend, with everyone else. Furthermore, a very important difference between the self in the online world and the Confucian relational self is that the latter cannot choose to be in the situation that she happens to be in, as previously discussed earlier. On the contrary, one can always choose to one's heart's content how one appears, what kind of friends one has, what kind of situation (what kind of social groups, etc.) one is in, and much else. This seems to show that the Confucian model might not be the most appropriate one for analyzing the online self. For the Confucian, one cannot of course choose one's own parents or one's siblings, one has to accept what Heidegger calls "thrownness" *(Geworfenheit),* the situation where one just finds oneself in without one's own design or choosing (Heidegger 2010). For Heidegger, one finds oneself "thrown" into one's own situated context in an existential sense. That is, one cannot but find that one is always in a situated context whereby one is aware of one's subjective condition vis-à-vis the alien world that confronts her. One is, in other words, "thrown" into this existential condition. In the Confucian perspective, on the other hand, one finds oneself in a web of relations with others and the Confucian does not have time or inclination to ponder the existential question of who one actually is in contrast to all these social conditions. The consciousness of there being oneself who is alone and distinct from all other selves one finds oneself interacting with does not primarily arise for the Confucian. Instead of pondering about one's existential condition, the Confucian self finds the custom of her community and her own historical tradition to be her own defining characteristic of her identity, characteristic which certainly involves her roles vis-à-vis other selves in the community. In the online world, this translates perhaps into the situation where the web of social relations in the online world are already established in which everyone in the world there has a clear role to play, as if everybody did not choose what role they were to play. This, of course, is contrary to the normal situation in the online world where everybody gets to choose, radically enough, who one is and what one wants to do.

2.9   The Self in Indian Philosophy

Compared with Chinese philosophy, Indian philosophy is much more akin to Western philosophy. There are an abundance of views on the metaphysical nature of everything including the self and the main question in Indian philosophy is the same as in Greek: What is it to be something? What is the essential nature of a thing? So coming back from Chinese culture most of us are on a familiar ground. The basic question of Indian philosophy is how to achieve salvation, and according to many strands of it this question is only answerable if one pays attention to the role of knowledge and wisdom. It is through the latter that salvation is achieved, a point of view that would not sound too strange to the ears of a typical Greek philosopher. Roughly speaking, there are two strands of thought regarding the self in Indian philosophy. One believes in the existence of a persistent self which endures even though its bodily cover is destroyed. This can be regarded as the mainstream view since it still is the dominant view in most Indian philosophies today, including some not so mainstream ones such as Jainism. According to these strands, the individual self exists in the form of an individual soul. For orthodox Hindu theory such as Advaita Vedanta, the individual self or soul is an *ātman* (Sanskrit for 'self'), an individual self, which ultimately is part of the larger cosmic self or the *brahman*. A usual comparison is with a drop of water and the ocean. An individual *ātman* is compared to a drop of water, which is essentially of one and the same nature with the *brahman*, which is compared to the ocean. As the drop of water dissolves into the ocean, so the individual *ātman* dissolves into the *brahman*. The ātman undoubtedly exists; this is the point of contention between Advaita Vedanta and Buddhism, as we shall shortly see later on. But it has one and the same nature as the ocean in the same way as a drop of water is of one and the same nature — that of water — with the ocean. Furthermore, another strand of Indian thought, that of Jainism, dispenses of the idea of the *brahman* altogether. Instead the *ātman* in Jainism, which is called jiva, stays individual and isolated from one another for eternity, each *ātman* being different from one another and retaining all the properties that qualify one as a particular individual the jiva once resided when the individual human being was still alive. We see then that no matter whether the *ātman* is ultimately dissolved into the one cosmic self or not, it has its own existence; it is there from the beginning.

The other strand of Indian philosophy regarding the self is a radical one in that it denies the ultimate existence of the self from the beginning. Buddhism is unique in all of Indian philosophical traditions in maintaining this view, and is perhaps in all the philosophical traditions of the world in this regard. In the remaining pages of the chapter I will outline in more detail the Buddhist view on the self, a subject which has given rise to much interest as it accords with the current critique of the idea of the self in contemporary philosophy in the West starting from Nietzsche onward. To be sure, Buddhist philosophy does not deny outright that the self exists. To do that would be insane, for we all know full well that we all have a self, something we always talk about whenever we use the first-person pronoun in our languages. However, what Buddhism denies is the view that what we refer to through the use of the word 'I' is ultimately speaking only a conglomeration of various factors which are held together only when various causes and conditions obtain. This is a complicated topic to which we now turn.

The first instance of the Buddha's teaching on the non-existence of the self is in the *Anatta Lakkhana Sutta* (Discourse on the Non-Self), which the Buddha gave to his first five disciplines soon after he attained Enlightenment. This is an important teaching, which helps us a lot in understanding the early Buddhist viewpoint on the non-self, so I am quoting it to the full:

> Thus I heard. On one occasion the Blessed One was living at Benares, in the Deer Park at Isipatana (the Resort of Seers). There he addressed the bhikkhus of the group of five: "Bhikkhus." — "Venerable sir," they replied. The Blessed One said this.
>
> "Bhikkhus, form is not-self. Were form self, then this form would not lead to affliction, and one could have it of form: 'Let my form be thus, let my form be not thus.' And since form is not-self, so it leads to affliction, and none can have it of form: 'Let my form be thus, let my form be not thus.'
>
> ...
>
> "Bhikkhus, how do you conceive it: is form permanent or impermanent?" — "Impermanent, venerable Sir." — "Now is what is impermanent painful or pleasant?" — "Painful, venerable Sir." — "Now is what is impermanent, what is painful since subject to change, fit to be regarded thus: 'This is mine, this is I, this is my self'"? — "No, venerable sir."
>
> ...
>
> "So, bhikkhus any kind of form whatever, whether past, future or presently arisen, whether gross or subtle, whether in oneself or external, whether inferior or superior, whether far or near, must with right understanding how it is, be regarded thus: 'This is not mine, this is not I, this is not myself.'
>
> (Anatta-lakkhana Sutta 2014).

Before discussing this Sutta, it has to be made clear that according to Buddhism the self is thought to be composed of five factors, namely form *(rūpa)*, sensation *(vedanā)*, perception *(saññā)*, thought formation *(sankhāra)*, and consciousness *(viññāna)*, which are called the Five Aggregates *(khandha)*. What this means is that the self is always analyzed as composed of these five aggregates and nothing else. 'Form' here is rather a technical term meaning the material part of the self, or the body, and the other four aggregates refer to the mental part of the self. The list of five elements here is intended to be exhaustive; there is no more to any conception of a self apart from these five and these five, taken together, are always necessary. The idea of the Sutta is that each of these five aggregates is examined and found not to be the self because if it were the self it would have possessed certain characteristics that a self must have, but it in fact does not. For example, if form or my body were to be my self, I should be able to control it, but the fact is that I can do it to be certain extent only. I can control my fingers to type on the keyboard at this moment when I am writing this chapter. But I cannot tell my fingers not to be painful when it is injured, and if I want it to be shorter or longer, or more proficient at the computer keyboard or at the piano, I am not able to improve their behavior just by telling them. In the same vein, my body incessantly grows older every minute; this is something I cannot arrest just by telling it to do so. On the contrary, if the body were really my own self, I should be able to control it completely, just as the body were indeed my own self, i.e., the one who is thinking at this moment and is telling the fingers, unconsciously to be sure, to type all these letters on the keyboard. The fact that things do not always follow what I want them to be, according to the Sutta, shows that things are not my self. Furthermore, when we analyze the mental components which are supposed to be part of the self—sensation, perception, thought formation, and consciousness—we find that they cannot be fully controlled either.

The mental part of the self always consists of these four elements; whatever we think or feel, desire, plan, whatever we do with our minds fall into one of these four types. However, we cannot fully control our mental components either. This is easily demonstrated by trying to focus your mind on something, thinking of only that thing and nothing else. After only a short while we will find that our minds wander off and cannot stay thinking of this particular thing for long. Examples like these show, according to the Sutta, that whatever we take to be our self, our very own identity, the who-we-are, is nonetheless devoid of the essential feature that we must take in order for them really to be our selves. This is basically the key content of the Doctrine of Non-Self—whatever we take to be components of our selves turn out not to be so, and no matter how extensive we search, anything that we come up with as putative component of the real self turns out not to be so. *Nothing* could count as our self.

However, if nothing could count as a self, then what it is that we refer to when we use the word 'I' and when somebody calls us by name, who is the one who answers? To this question I would like to quote a passage from *The Questions of Milinda,* a text containing a series of dialogs between a monk named Nāgasena and King Milinda, who presumable was a Greek whose actual name was Menander. According to the tradition, he was a descendant of one of Alexander the Great's generals who stayed on in India after the latter's foray into the area. The passage here is one of the most famous passages in early Buddhism that deals with the question of the self:

> Now Milinda the king went up to where the venerable Nāgasena was, and addressed him with the greetings and compliments of friendship and courtesy, and took his seat respectfully apart. And Nāgasena reciprocated his courtesy, so that the heart of the king was propitiated.
>
> And Milinda began by asking, 'How is your Reverence known, and what, Sir, is your name?'
>
> 'I am known as Nāgasena, O king, and it is by that name that my brethren in the faith address me. But although parents, O king, give such a name as Nāgasena, or Surasena, or Virasena, or Sihasena, yet this, Sire, –Nāgasena and so on–is only a generally understood term, a designation in common use. For there is no permanent individuality (no soul) involved in the matter.'
>
> (The Questions of King Milinda 2014).

Nāgasena starts by telling Milinda that there is actually no "permanent individuality" or "no soul," and Milinda then raises a series of questions designed to challenge this view. Starting by pointing out the dire consequences of there not being Nāgasena, such as that there would be no one who is responsible for crimes or no one who will reap the reward of good deeds, Milinda then raises a series of questions designed to show that what is understood to be Nāgasena is in fact an illusion. These questions are intended to confound Nāgasena because earlier the latter admits that his brethren in the temple call him by that name, implying of course that there be someone who bears the name. Milinda asks his interlocutor to point to various parts composing what is thought to be the person or the self of Nāgasena and then asks whether those parts are the sought after self or not. Pointing at the leg, for example, the question is whether this organ, this thing being pointed at, the person of Nāgasena. Nāgasena then answers in the negative. The same strategy is then applied to all the parts that comprise Nāgasena's body and the answer is the same. Then the same strategy is applied to the mental components of the Five Aggregates (*skandhas* in Sanskrit or *khandas* in Pali). Pointing at this thought or this sensation, the question is asked whether this particular thought is the person, and the answer is obviously no. Repeating the process yields a conclusion that none of the various mental components is the person either. Having exhausted all the components of the self, Milinda then declares that there is in fact no Nāgasena and that his words uttered earlier that there be a Nāgasena in the sense of who is answering his brethren's calls is declared to be false.

To this Nāgasena raises the famous example of the chariot. The chariot, as is well known, is also of various parts, but none of the parts taken individually is the chariot. Furthermore, when the parts are taken together, they still don't become the chariot because the parts have to be taken up only in a certain way. After satisfying the King that the chariot cannot be anything apart from these parts either, Nāgasena then concludes that there is no chariot, and that the King has rode on nothing when he came to see the monk. The King then replies that the chariot is just nothing but its various parts when they are assembled together in one particular way. Then Nāgasena replies that the self is the same. It is not that there is absolutely no self, but the self is what emerges from its various component parts when they are assembled together in one definite way. It is in this way that the designation 'Nāgasena' or 'chariot' applies in their respective cases.

2.9.1 Buddhist Philosophy and Online Self

Our import in understanding the online self is that we also understand the online self to be composed of its component parts in the same way. There is, in one manner of speaking, no self, online or otherwise, but in another manner the self is there as it emerges when the parts are assembled, just like the chariot emerges when its parts are assembled in the correct way. We have to remember that this way of understanding the self means that there is no core to the self. That is, there is no self subsisting soul which underpins the self's putatively real identity. This is the key difference between the Buddhist on the one hand and the Hindu and Jain conceptions on the other. That there is no core to the self does not imply that the self does not exist. Nāgasena the monk does exist even though when pointed to each of his part none of it is one and the same as the monk; nonetheless the monk himself does exist. In the same vein, even though none of the component parts making up a self—bodily and mental episodes—is one and the same as that self, when the parts are assembled in such and such a way, then according to common designation the self does emerge. Even what is normally considered to be the soul, the core identity of a self, is itself composed of various parts and can be analyzed in the same way. Kantian and Lockean consciousness of the self and personal identity, according to the Buddhist, can also be analyzed in this way too, which results in the realization that there is in fact no substantive, independently existing self.

In the same vein, the self in the online world is also subject to the same analysis. Its emergence as a self is due to some kind of designation or conceptualization: Various component parts that altogether make up the identity of the online self are 'taken up,' so to speak, in the process of designation or conceptualization and, assembled in such a such a way, these component parts are taken together to belong to a new entity, in this case a self or a person. That the online self does depend for its existence on its recognition and designation as such by outside factors is very important, and is a key issue in my argument for the externalist conception of self and identity later on in the book.

It might be asked how the Buddhist conception respond to Kant's view that the 'I think,' namely the Unity of Transcendental Apperception, accompanies all my representations such that they fall into being *my* mental episodes? Recall that for Kant the 'I think' is the root of the consciousness of there being a unified self and functions as the condition of possibility of there being a coherent conception of the world (and of oneself) which is necessary for any form of knowledge in the first place. The key element of the 'I think' is that it is purely subjective—it never functions as an object in the sense of being thought of. However, according to the Buddhist conception we have seen so far, the analysis looks as if it goes toward the self as an object only. When the conglomeration of mental episodes that function together as a self are analyzed and examined objectively, the episodes are found not to possess the requisite element that would have qualified them to be one and the same with the personal self. The Buddhist, so it seems, does not examine the self qua subjective thinker. So where can the Buddhist analog to the Kantian 'I think' found? I think that the Buddhist, at least in the canonical text of the Tipitaka, which is considered to contain the Buddha's own original words, is rather silent on the issue. However, this is understandable because to posit a transcendental subject such as the 'I think' presupposes that there be a distinction between the subject (one who thinks) and the object (whatever is thought) from the beginning, so when the Buddhist tries to demolish the whole distinction they cannot rely on yet another distinction between the knower and the known existing at a deeper level. The Buddhist, in other words, cannot proceed by arguing that there be a knower, a subject, who shows that the self, examined objectively, does not consist of any element that qualifies it to be a self. This is understandable because the Buddhist rejects all kinds of inherently subsisting knower at any level (to posit a knower at any level would be tantamount to positing a substantive subject, which is in contrary to the Buddhist spirit). So the analysis that one finds in texts such as *Anatta Lakkhana Sutta* or *Milinda Panhā* cannot presuppose such a transcendental knower. The strategy is that for any knower at any level, they are all subject to the same kind of analysis offered here, and to presuppose that there be a higher subject would be to succumb to the preconceived, unexamined view that there has to be a substantive subject for all kind of mental operations. According to the Buddhist, there can be a subject for a mental operation, but such subject does not have to be one and the same throughout. There can be neither evidence nor argument in support of that claim.

2.10   The Fragmented Self: East and West

The Buddhist critique of the inherently subsisting self finds an unlikely counterpart in the philosophical movement of the critique of the self that started in earnest in the nineteenth century. In the eighteenth century Hume presented a view that is rather similar to the Buddhist in that he proposes to do away with the existence of the self altogether. Arguing in the usual way of basing everything on impressions and ideas, which only have their original source in the sense experience, Hume proceeds to argue that the self as commonly understood does not exactly exist because what do exist, as ideas and impressions, are those that are immediately perceived, such as various mental episodes and so on. Since none of those constitute the self, Hume then concludes that the concept of the self belongs to the same type as the concept of causation; that is, they are constructed out of the mind's habit when perceiving certain regularities. Let us recall that for Hume the concept of causation cannot be found as an impression directly perceived by the mind through sense experience. On the contrary it is an idea which is derived from a certain number of impressions, especially those that are regularly connected with one another through, in Hume's words, the relation of "constant conjunction." In the same vein, episodes in the mind that result from the mind's own introspection into its own goings on reveal that they are also connected through habit and custom into a coherent self. But since the idea of the coherent self is based on no more than this habit and custom which means that they are not necessarily connected with one another, Hume concludes that the self does not exactly exist because if it were to exist there must have been some kind of necessary connection between the impressions (See Hume 1966, especially *A Treatise of Human Nature,* Book I, Part IV, Section VI, pp. 256–271). However, this does not imply that for Hume there is absolutely no self at all; to say that would be little more than crazy because one refers to one's own self all the time one uses the first person pronoun 'I'. As the idea of causation is based on no firmer foundation that habit and custom formed through perception of constant conjunction, so too the concept of a self is formed through habit of seeing regular connections—such as one belonging to the same first-person pronoun through its various uses or the same consciousness—which also implies that the self does in a sense exist though not on the firm foundation of sense experience.

Hume's view here finds a sympathetic reverberation in Buddhist philosophy (See, for example, Giles 1993; Siderits 1997; Tomhave 2010). We have seen that the basic Buddhist strategy is to analyze the self into various components and examine each of them in turn, and only when these components are put together in one particular way does the self emerge, which shows that the self is a kind of a constructed entity and does not exist in its own right. Hume does not talk about construction of the self in this way, but he maintains that the idea of the self arises out of habit and custom, as we have seen. The two philosophical traditions have their own way of maintaining the putative existence of the self, each in its own way.

Furthermore, in the nineteenth century some Western philosophers started seriously to question the rationality on which the modern conception of the self is based. Philosophers such as Schopenhauer, Kierkegaard and Nietzsche attacked the idea of the self as the seat of rationality and pure agency, and started to see the self more as a victim of circumstances in the sense that it is a vehicle of forces stronger than itself which it cannot resist. Schopenhauer introduces the concept of the blind and powerful Will, which functions through the individual in such a way that the latter cannot resist since the latter is constituted by the Will itself. Kierkegaard directly repudiates the role of rational agency of the self, preferring instead of talk about 'leap of faith,' a disposition which involves directly and consciously rejects reason altogether in order return to one's original existential condition. Nietzsche finds rationality to be only a veneer beyond the Will to Power, a means by which individuals gain power over whatever they would like to overcome, which in this case involves cunning and deceit because stronger weapons are not available to them, prompting them to use reason to gain power over the physically stronger. There are obviously much more details in the thoughts of these three philosophers than can be summarized here. The point is merely that these three philosophers strongly criticize the idea, one that has been dominant since the time of the Greeks, that the self is the seat of rationality and freedom. These critiques of the self then form a basis for the more thorough critique of the self as well as of rationality and objective knowledge in general in the so-called 'postmodern' thinkers. There is not enough space in this book to discuss on these interesting topics in any detail; however, I would like to point out that, no matter how much Schopenhauer, Kierkegaard and Nietzsche criticize the idea of the rational and free self, they do not deny that the self exists. It's perhaps not their aim to deny the self, but they instead aim at criticizing rationality and the idea that the self is constituted by reason and freedom.

2.11 Conclusion

We have seen, then, a very brief history of the self in both the West and the East. We find that Buddhism is alone among all the philosophies discussed so far (with a possible exception of Hume) that denies the ultimate existence of the self. That is, Schopenhauer, Kierkegaard and Nietzsche more or less accept that the self exists in one form or another, and they do not bother to expend their energies into refuting the view that the self does exist. Their aims, as we have seen, involve rather presenting a critique of rationality, especially the view that the individual self is the vehicle through which pure rationality emerges, something we have seen to be most visible in Kant. Nonetheless, when we compare the critique of the rational self by these three philosophers with Buddhism, we find that neither of these philosophers denies the existence of the self. However, there are some interesting points of comparison. It is well known that Schopenhauer was very interested in Indian philosophy and

Buddhism, and cited many passages from Indian philosophical texts, some of which had become available in his time. Schopenhauer usually mentions Buddhism when he talks about how to overcome the force of the Will, which for him is only possible through the ascetic disposition of denial of the self. Since the self in its normal condition is always slave to the Will, the only way to gain ultimate freedom, to be free from the clutch of the Will, is through transcend it through art, music, and ultimately asceticism (Schopenhauer 1964). Since the self is a product of the Principle of Individuation *(principia individuationis)*, which is an expression of the working of the Will when it manifests itself in the empirical world, asceticism culminates in denial of the self altogether (See, for example, Hongladarom 2014). What are interesting to us are then the idea that in Schopenhauer we find a conflict, an *agon*, between the Will on the one hand and the individual self on the other, which shows that the self itself is something that is capable of resisting the very environment in which it finds itself. This is certainly not a Buddhist point because the way toward ultimate release from suffering involves realization and knowledge, especially the realization that, ultimately speaking, the self and its environment is one and the same.

With this brief history of the self in both the East and the West, it would be very interesting to see how an analysis of the online self reveals what kind of entity it is, or whether it could be an entity at all. However, what these stories of the self reveal is that they are analyses of the self in the so-called "offline" mode. This is perfectly understandable because none of the philosophers discussed here is alive in the age of the Internet. Nonetheless, we could see that the online self is a kind of a projection of the self into the public world, and this has been around since humans started to become self-aware. A big problem for us is what kind of relation obtains between the normal "offline" mode and the presented mode of the self, which includes how we present ourselves to the world in the traditional and the online modes. Key to this point must be a philosophical analysis of the metaphysical status of the self itself, which we have discussed so far in the chapter. Thus we could also look at our history of the self presented here as a preamble, and the history of the self seems to enter a very intriguing chapter when the self enters the age of Facebook and Twitter.

References

Anatta-lakkhana Sutta: The discourse on the not-self characteristic. (2014) Translated from the Pali by Ñanamoli Thera. Retrieved from http://www.accesstoinsight.org/tipitaka/sn/sn22/sn22.059.nymo.html

Clark, A., & Chalmers, D. J. (1998). The extended mind. *Analysis, 58,* 7–19.

de Spinoza, B. (1985). *The collected works of Spinoza, Volume I.* (E. Curley, Trans., Ed.). Princeton: Princeton University Press.

Ess, C. (2005). "Lost in translation?": Intercultural dialogues on privacy and information ethics. *Ethics and Information Technology, 7*(1), 1–6.

Ess, C. (2006). Ethical pluralism and global information ethics. *Ethics and Information Technology, 8*(4), 215–226.

Ess, C. (2007). Bridging cultures: Theoretical and practical approaches to unity and diversity online. *International Journal of Technology and Human Interaction, 3*(3), iii–x.

Ess, C. (2010). The embodied self in a digital age: Possibilities, risks, and prospects for a pluralistic (democratic/liberal) future? *Nordicom Information, 2–3*, 105–118.

Ganeri, J. (2012). *The self: Naturalism, consciousness, and the first-person stance.* New York: Oxford University Press.

Geiger, I. (2007). *The founding act of modern ethical life: Hegel's critique of Kant's moral and political philosophy.* Palo Alto: Stanford University Press.

Giles, J. (1993). The no-self theory: Hume, Buddhism, and personal identity. *Philosophy East and West, 43*, 175–200.

Goffman, E. (1959). *The presentation of self in everyday life.* New York: Anchor.

Heidegger, M. (1977). The question concerning technology. In *Basic writings* (pp. 283–318). New York: HarperCollins.

Heidegger, M. (2010). *Being and time.* (J. Stambaugh, & D. J. Schmidt, Trans., Rev.). Albany: State University of New York Press.

Hongladarom, S. (2014). Schopenhauers Metaphysik des Willens und Nagarjunas Konzept der Leere. In H. Detering, M. Ermisch, & P. Watanangura (Eds.), *Der Buddha in der deutschen Dichtung* (pp. 39–50). Göttingen: Wallstein Verlag.

Hume, D. (1966). *Selections from an enquiry concerning human understanding and a Treatise of Human Nature.* La Salle: Open Court.

Kant, I. (1929). *Critique of pure reason.* (N. K. Smith, Trans., Ed.). London: St. Martin's.

Kant, I. (2012). *Groundwork of the metaphysics of morals.* (M. Gregor, & Timmermann, J., Trans., Eds.). Cambridge University Press.

Korsgaard, C. (1996). *The sources of normativity.* Cambridge/New York: Cambridge University Press.

Kurzweil, R. (2005). *The singularity is near: When humans transcends biology.* New York: Viking.

Locke, J. (1997). *An essay concerning human understanding.* In R. Woolhouse (Ed.). New York: Penguin.

Martin, R., & Barresi, J. (2006). *The rise and fall of soul and self.* New York: Columbia University Press.

Owens, J. (1988). The self in Aristotle. *The Review of Metaphysics, 41*(4), 707–722.

Rawls, J. (1971). *A theory of justice.* Rev. Ed. Cambridge, MA: Harvard University Press.

Rosemont, H., Jr. (1991). Rights-bearing individuals and role-bearing persons. In M. I. Bockover (Ed.), *Rules, rituals, and responsibility: Essays dedicated to Herbert Fingarette* (pp. 71–101). La Salle: Open Court.

Schopenhauer, A. (1964). *The world as will and idea, Vol. 1* (K. B. Haldane, & J. Kemp, Trans.). London: Routledge.

Searle, J. (1997). *The construction of social reality.* New York: Free Press.

Shields, C. (2009). The Aristotelian Psuchê. In G. Anagnostopoulos (Ed.), *A companion to Aristotle* (pp. 292–309). Chichester/Malden: Blackwell.

Siderits, M. (1997). Buddhist reductionism. *Philosophy East and West, 47*(4), 455–478.

Siderits, M., Thompson, E., & Zahavi, D. (Eds.). (2011). *Self, no self?: Perspectives from analytical, phenomenological, and Indian traditions.* New York: Oxford University Press.

Taylor, C. (1989). *Sources of the self: The making of the modern identity.* Cambridge, MA: Harvard University Press.

The Questions of King Milinda. (2014). (T. W. R. Davis, Trans.). Retrieved from http://www.sacred-texts.com/bud/sbe35/sbe3504.htm

Tomhave, A. (2010). Cartesian intuitions, Humean puzzles, and the Buddhist conception of the self. *Philosophy East and West, 60*(4), 442–454.

Chapter 3
The Extended Self View

In the last chapter we have seen a brief overview of the history of the self as it wended through history both in the East and the West. This chapter will offer a philosophical analysis of the self. The view to be presented and argued for is called "externalist" theory of personal identity, which I would like to call for short the Extended Self View. The idea is that criteria for identifying a person through time have to come from outside of the subject's introspective mental content. External factors must play a formative role in the subject's or the person's identity through time. As for identity at a time external factors play the role too. Basically put, the idea advanced here is that an identity of a person, the fact that he or she is the same person through time or at a time, is constituted through external factors. For example, since memory is usually fallible, one often has to rely on devices such as written records or testimony of others to corroborate the memory. Cases of failing memory or confusion between imagination and memory (where a scenario from the past cannot be distinguished as memory or imagination) seem to show that memory is not entirely reliable. This seems to require the aid of the external factors such as the notebook, diary and so on. A story that one builds up to tell oneself who he or she is, one's life narrative so to speak, is little more than a construction that one makes up in order to make sense of who he or she is. In this sense the narrative is little different from the aids such as notebooks and diaries in that they make up the sense of who one is. Furthermore, the sense of who one actually is appears to be constituted also by the story and the perception that others have of the person also. The sense of who I actually am seems to require input from my relatives, friends and colleagues, those who know me and who have a perspective of who I am from their own individual point of view. Who I actually am is not only a function of what I make of myself. Sartre's dictum that "existence precedes essence" thus does not seem to ring true here. (Nor is it true that essence precedes existence either. This is because, as I shall argue, the whole idea of there being any essence of any object is suspect from the beginning. On the contrary, both essence and existence seem to construct each other, so they arise at the same time.). I will spend the rest of the chapter arguing for all these points.

© Springer International Publishing Switzerland 2016

S. Hongladarom, *The Online Self*, Philosophy of Engineering and Technology 25,
DOI 10.1007/978-3-319-39075-8_3

Another related point that will be argued for in some detail in this chapter is the view that the self is informational in nature. That is, a self is constituted by information rather than by information rather than merely by flesh and blood or by successive mental contents. For example, the facts that I am a lecturer at a Thai university, that I am known as such and such by my relatives and friends, and so on, are items of information that make up who I am. Hence my view is quite similar in some respect to that of Luciano Floridi. However, I will spell out my differences from Floridi. Basically speaking, the difference is that, while Floridi focuses on his metaphysical analysis of the informational nature of everything, including the person (For example, Floridi 2005, 2006, 2011a, b). I focus more on the self, which is an elusive object whose nature and existence does not seem to be on a par with any concrete object. The most salient difference between my position and Floridi's seems to be focused on the ontology of the self. While Floridi maintains that narrativity or internal coherence is sufficient for personal identity, I believe that it is not enough.

So here is a brief outline of this chapter. I will start out with an introduction of the problem of personal identity. This is mainly for the benefits of students and those who are relatively new to the field. In the next section I will present the main argument of the chapter, which is to argue for the externalist position of personal identity. This position is inspired by Buddhist philosophy, and I shall make it plain in the course of the chapter why it is so. The main position and argument of Buddhist philosophy on this topic will be given; this is not merely a rehash of what has been discussed quite extensively in the last chapter, but in this chapter I will present my own argument as to why a position that is akin to the Buddhist one should be the most tenable position among those that are available for the problem of personal identity. The position I will argue for in the chapter will not be a purely Buddhist one—after all the Buddha does not specifically address the problem of personal identity as it is being discussed by modern philosophers. (In fact the Buddha has a position, which is a non-position. That is, he maintains that the ultimate view is that all views are to be relinquished.) But in any case it is derived from the spirit of the Buddhist position.

Then in the next section I will argue for the externalist position in some detail. This will proceed by showing that its main competitor, the internalist position that bases criterion for identity on introspectible content, is not as tenable as the externalist one. Furthermore, the view that bases personal identity on bodily continuity will also be shown to be untenable. In a way physicalist theories are internalist too because the bodies that are believed to belong to the person are somehow internal to the person herself because it is after all her body. Afterwards the implication of the externalist theory on an understanding of the online self will be given in detail. The idea is that, as I shall argue, the externalist theory appears to be the most cogent and effective one in helping us understand the issue. As personal identity is constituted through external factors (such as memory aids, testimony, and so on), the same applies for analyzing the identity of the online self. A key point to be advanced and argued for is that there is a continuity between the self as it exists in the "offline" and online world. This continuity is not only due to the obvious fact that the offline self is the one who manipulates and constructs his or her online persona, but the continuity here is also metaphysical. That is, there is a metaphysical connection between the two selves, in the same way as there is a connection between my own self as existed, say, 2 months ago and as it exists right now as I am typing this. In the one case there is a temporal dimension—one that separates my two-month-ago self and my self right now; in the other case, there is a spatial and ontological dimension that functions in the same way with the temporal dimension in that they distinguishes episodes of my self, although in different ways.

The foregoing sections form the first part of the chapter. In the latter part I will discuss a related but different topic, that the self is informational in nature. This means that the self is constituted through information. This does not have to mean that the self is totally reducible to information, as if the body, the flesh and blood, or the personality and mental life of the person does not matter, and the self or the person is just some ethereal information flowing through the wire or the air in bits. Perhaps this is my main difference from the view of Floridi. My ultimate position, inspired by Buddhism, is that the self has no ultimate nature, because it is a construction from the beginning. Floridi, on the other hand, seems to hold that information is such an ultimate nature. Nevertheless, I agree with him that to analyze the self as information seems to help us understand it better, especially when we analyze the self in relation with its online counterpart.

3.1   Problem of Personal Identity

The main point of the problem of personal identity is what accounts for the identity of a person as he or she seems to persist through time. That the person seems to so persist is not much of a problem since it is rather obvious. I am certain that I am the same person a day or two before as the one who is typing this text at this moment. However, that I am certain is one thing, how I can find evidence to illustrate and justify that claim is another. As we have seen from the Buddhist position in the last chapter, simply to claim that my body 2 days ago is the same as my body right now does not work because it is well known that cells and other materials inside the body are always changing. I was a bit lighter 2 days ago than I am now because as of this moment I just had a rather big lunch. Some cells in my body have quite a short life. The red blood cells, for example, live only for a few days and are constantly replenished by the body. Other cells, though they might live a bit longer, do suffer the same fate. If my body is composed of nothing but cells, then it seems that my body is continually changing and being recycled; in this sense my body is more like an event than a static thing.

Alternatively, I might identify who I am which persists through time with my personality. I have certain memories, certain sense of who I am that is distinct from being other people, some sense of uniqueness that I believe I possess that makes me distinct from other people. This sense of uniqueness resides perhaps in my mental stream, but when we fully analyze that stream we find, as Hume does, a series of continually changing episodes none of which stay long enough to qualify as a basis for my persisting as a distinct personality. All this shows that personal identity is deeply problematic. On the one hand we seem to be certain that we are some kind of entities that persist through time; on the other, we just cannot seem to find any basis for such persistence. Neither bodily or mental continuity appears to count. The "continuity" in question seems to be the same kind as a flowing river is continuous. As a river is constantly flowing yet remains the same river; our person is the same even though everything in it is constantly flowing. If we can't find any such basis, then we have to admit that personal identity is utterly mysterious, or worse that we don't have any identity at all. Either choice is not acceptable, not least because our identity is so obvious to us.

The philosopher who first raised the problem of personal identity as we know it is Locke. According to Locke, there is a problem of accounting for one's being the same person from one moment to another. For example, when I was 2 years old, my body then was very different from the aging body of a 51-year-old college professor who is now typing this text, so what accounts for their being the body or a part of one and the same person? Locke's answer, as is well known, is that my memory serves to link up the two-year-old boy with the 51-year-old. I introspect my life history and go back to the episodes of my being a young toddler (so far as I can recall) and this chain of memory serves to provide a basis of my being one and the same person through time. This is known in the literature as the Psychological Account, and is one of the main theories of personal identity today.

The basic idea of the Psychological Account is that my mental episodes, i.e., my thoughts, memories, musings, feelings, etc. make up who I am, and those who share the same set of thoughts and feelings would essentially be me. According to Derek Parfit, personal identity is reducible to psychological continuity. In a thought experiment where my brain is split into two and implanted into two bodies, Parfit holds that there is a sense in which 'I' survive in these two bodies (Parfit 1971). For Parfit the difficulty of this dilemma (how could there be two "I's"?) stem from the underlying belief that personal identity must be unitary—to contemplate there being two I's, two sets of subject experiences both of which are "me"—just does not seem to make sense. However, if we get rid of the underlying belief and accept that there is a sense in which there can be two "me's" in the same way as there can be two groups which are separated from one original group. In this way Parfit's argument looks rather similar to the Buddhist. In any case, for Parfit identity of a person just consists in his or her psychological continuity, and the continuity here can certainly admit of degree. Thus, my person 1 hour ago is more connected to the state of myself as of now than, say, my person 1 day or a week ago (See Parfit 1984, and also the interview of Parfit in Pyle 1999). Many philosophers differ from Parfit here, though they still subscribe to the Psychological Account.

Another major theory of personal identity could be called the Bodily Account. Instead of psychological continuity, this theory looks at continuity of the body rather than the mind as basis of identity. Thus, Bernard Williams or Eric Olson would have it that an organ such as the brain or the whole body is all there is to personal identity. In fact Olson, in his Biological Account, maintains that it is the whole of an organism, a human, or a cat, that is responsible for personal identity, and he argues that his account is different from the physicalist account in this regard (Olson 2007, 2011). However, for our purpose we could group these two broad accounts together, and put the physicalist account, one that maintains that the brain alone is responsible for personal identity, as a species of the Bodily Account.

To illustrate what the Bodily Account, broadly speaking has to say, we look again at a brain transplant thought experiment (it has to be a thought experiment now because the technology is advanced enough actually to transplant the brain). If A's brain is put into B's body (whose brain has been removed), then according to the Bodily Account it is still the person B that survives, not A. It is the body that counts, not just one organ such as the brain. Certainly a human body undergoes a lot of changes — it has to eat, secrete, breathe, and so on, but the point of the Bodily Account is that, within the frame of this particular body lies the seat of personal identity. I can remain the same person just because I happen to be this particular body that inhabits this location in space and time. One might think of the Ship of Perseus, which is replaced a little by little, until at the end not one part of it remains the same. In this case we still call it the Ship of Theseus because it still looks like the original ship. However, the most important reason is that the ship still occupies the same spatiotemporal location. Even though all its parts have been changed, the overall structure that consists of those parts still remains the same. Even if the ship changes its shape somewhat (but not too much as not to remain a ship) we still call it the same ship simply because the conjoined parts are there — it's not like the parts are scattered everywhere which would mean that the ship does not survive. In the same way, then, the fact that all of my cells are replaced after a period of time does not imply that I do not remain the same person.

Another account of personal identity that has recently gained an increasing number of adherents is the Narrative Account, such as Marya Schechtman and Luciano Floridi, among others (Schechtman 2007, 2012; Floridi 2011a). According to this account, identity of the person is maintained through their role in a narrative: If the narrative is coherent and plausible, then the identity of the person who plays a role there is maintained. There is an obvious problem with this account, which is similar to the Memory Account. What is the necessary factor that maintains coherence of a narrative? Suppose there is a story of someone that tells her biography from her birth up until her death. What we would need if the Narrative Account were to be successful as an account that supplies personal identity is that the narrative or the life story alone would account for the identity of the lead actress in the narrative. What we need, in other words, is that the narrative or the story is sufficient and necessary in maintaining that the person being narrated about her identity. But it would seem, on the contrary, that the main reason why the story is a coherent one in the first place is that the protagonist of the story is one and the same person through time from the beginning. There is a biography of someone; the reason why that biography is a coherent story seems to be because it is a story of some identical person from one time (for example her birth) to another (her death). The biography would not be what we normally expect of life stories if it were to be a story of two persons. The "biography" here can be fully coherent — that is the point — but the fact that it is a coherent story does not seem to imply necessarily that it is thereby a story of one and the same person. On the contrary, that the person is one and the same appears instead to be the factor that results in the story being coherent. But if this is the case, then the Narrative Account does not seem to be doing the job that it has been designed to do.

3.2   Criticisms

It's my contention that neither the Psychological Account, the Biological Account, nor the Narrative Account is ultimately tenable when it comes to personal identity. This is because, ultimately speaking, there can be no identity of a person when we look at the issue from the inside, either inside the physical constitution, the mental episodes, or the narrative account of the person. This does not of course mean that there is no identity at all. To hold that view would be clearly irrational, if not downright crazy. However, I will argue that when we account for factors responsible for our identity as the same person through time we cannot look solely at our psychological or bodily continuity. Such factors, then, have to lie elsewhere.

3.2.1   Against the Psychological Account

Among the various views that comprise the Psychological Account, the most obvious one is one that, following Locke, takes memory as the criterion of personal identity. However, a standard objection against memory has been introduced since the eighteenth century. Joseph Butler (2008) argues that the memory account is circular, because in order for me to be certain, when I inspect my memories in order to find out how I have become the same person, those memories have to be mine from the beginning (otherwise I would have no clue to ascertain to myself that those are in fact my memories), but the only way for me to be certain that those are my memories is that I have to be able to remember them. So the only way I can rely on my memory is that I have to use my memory to do so. This is hardly a recipe for success. In Butler's words: "But though consciousness of what is past does thus ascertain our personal identity to ourselves, yet, to say that it makes personal identity, or is necessary to our being the same persons, is to say, that a person has not existed a single moment, nor done one action, but what he can remember; indeed none but what he reflects upon. And one should really think it self-evident, that consciousness of personal identity presupposes, and therefore cannot constitute, personal identity, any more than knowledge, in any other case, can constitute truth, which it presupposes" (Butler 2008, p. 100). In short, relying on memory as the criterion of personal identity is circular because using memory presupposes identity, but the idea is that identity is constituted by memory.

Furthermore, Thomas Reid added a few more objections to the Lockean account (Reid 2008, pp. 113–118). According to Reid, if a person at stage t1 in his life experiences something called e1, and the same person at stage t2 experiences e2, and this same person again at stage t3 experiences e3. Suppose that the person at t3 is quite advanced in age and only remembers stage t2, but not t1, but the same person at stage t2 can remember t1 well. Then according to Locke's account there have to be two persons because it is memory or consciousness of being the same person that is the criterion, but the hypothesis is that the person here is the same throughout, so for Reid Locke's account lands him in a straightforward contradiction.

Reid also has two more objections to Locke that deserves to be discussed in detail. The first one is that memory or consciousness is always changing and transitory; this is a normal fact of our consciousness. We simply cannot fix our thoughts so that we stay thinking about the same subject for any length of time. This transitory and ever changing character of consciousness makes it, according to Reid, a very poor criterion of identity, for it would mean that our identities would be transitory and ever changing too. The idea here is that memory is not very reliable; we all have experiences where we confuse genuine memory with imagination. We recall something in some detail and the deeper we delve into our memories the less certain we are whether that is a genuine memory or simply our imagination. We might be certain to ourselves of some strong episodes that made an indelible imprint on our brains, some very important moments in life that we cannot forget, such as the wedding day, the death of a parent, and so on. But when we try to relive those moments in further detail we find that we become less certain whether those finer details actually happened or not. This is just not the way our brains work. Memory is selective, and the more details we want to squeeze from our memories the more likely that what is supplied is instead our imagination. If this is the case, then memory is not entirely reliable. Reid's point that consciousness is always changing, then, implies that if we want to base our personal identity on something of such a transitory nature, then our identity would be transitory too: "Consciousness, and every kind of thought, are transient and momentary, and have no continued existence; and, therefore, if personal identity consisted in consciousness, it would certainly follow, that no man is the same person any two moments of his life" (Reid 2008, p. 116).

Reid has another objection to the memory account which is stronger than the transience objection. When one talks about things being identical or the same, one has to make clear whether those things are identical in the same as being one and same objects, or simply belonging to one and the same kind. For example, two coins of exactly the same kind, say two American 5-cent coins, are exactly identical in the sense that they are 5-cent coins of the same type. However, they are not identical in the sense of their numerical being; that is, they are simply different things and we can easily count them as two coins. In philosophical jargon one talks about two things being type-identical or token-identical. Things are type-identical when they belong to the same type, or made from the same mold so that they look exactly alike, but that by no means implies that they become one and the same thing. On the contrary, things are token-identical, or identical at the level of token, if they are exactly one and the same thing but perhaps merely called differently. In the famous story

Clark Kent is type-identical with Superman; it is not only that the two belong to the same kind, that of being a Kryptonite, but the two are exactly one and the same. Reid's point is that our episodes in our consciousness cannot be token-identical to any episode we might have in the past, and it is only token-identical episodes that suffice in bringing about identity. According to Reid,

> When Mr. Locke, therefore, speaks of "the same consciousness being continued through a succession of different substances"; when he speaks of "repeating the idea of a past action, with the same consciousness we had of it at the first," and of "the same consciousness extending to actions past and to come"; these expressions are to me unintelligible, unless he means not the same individual consciousness, but a consciousness that is similar, or of the same kind.
>
> If our personal identity consists in consciousness, as this consciousness cannot be the same individually any two moments, but only of the same kind, it would follow, that we are not for any two moments the same individual persons, but the same kind of persons" (Reid 2008, p. 117).

When I remember some event in the past, what is occurrent in my mind is an episode in at the present time, whose content is that of a time in the past. This episode then is only type-identical with the episode that I had in the past, and cannot be token-identical, so according to Reid, memory cannot be sufficient in bringing about personal identity.

Proponents of the memory account counter to these moves by Butler and Reid by loosening up the memory requirement. Memory perhaps is not sufficient, but psychological continuity should be. In this case I don't have actually to remember everything in order for the remembered episodes to be sufficient for my personal identity, the fact that my mental episodes can link back to the past is sufficient even if I might not remember all those episodes that happened in my life. This is the concept that Parfit relies on in his account of personal identity. The basic idea is that if there is a chain of mental events from the beginning of my conscious life until the present, then my personal identity consists in such a chain. In this case, if my brain happens to be transplanted onto a different body, then that new body would house my person, as if I can shed my own body and take on a new one just like a snake shedding its old skin. In Indian philosophy there is a usual talk of the soul transmigrating from one body in one lifetime to another "just like a person changing clothes." The idea is roughly the same, except that in talking about psychological continuity one does not have to posit the immortal soul, only the continuity of conscious states is enough. Functionalists who hold that the mind is essentially similar to a software also believe the same thing. It does not matter what hardware the program is running on, so long as it is the same software enabling the various hardware configurations to do the same functions then these configurations become essentially identical. One might make an analogy comparing the brain to a computer and the configurations of the firing and interacting neurons inside the brain with the software. If these same configurations are to be housed in another physical brain, then according to the functionalists that new brain would be the brain of the same person to whom the configurations belong. In the same vein, if one takes the brain to house the software and the neuron configurations, then if that particular physical brain with all the configurations remaining intact were to be implanted in a new physical body, then that body becomes that of the same old person. The reason why this arrangement gives an impression to some that there is only one person who merely changes his or her body is that they seem to subscribe to the idea that psychological continuity is all there is to personal identity. In these cases, the continuous mental stream functions essentially as the soul does in ancient philosophy. As the soul is the seat of personality and identity of the person, so does the continuous stream.

However, it is rather difficult to see exactly how the continuous stream of psychological continuity could actually function as something constitutive of personal identity. According to the Psychological Account, if two persons, namely P1 at the time t1 and P2 at t2, share this same continuous stream of psychological continuity, then P1 and P2 are the same person. However, is it possible that there be two persons who share roughly speaking the same stream of psychological continuity? That this scenario might sound counterintuitive to some might be only because to them the apparent fact of the matter is that psychological continuity is all there is to personal identity, so any scenario that two different persons share the same psychological continuity just does not arise. However, we can readily imagine a thought experiment where two different persons share the same psychological continuity, and if this is actually possible, then this appears to land the Psychological Account in a difficulty. Suppose P1 and P2 share roughly the same psychological continuity. This alone does not collapse the two persons so they become one and the same. In the recent film, Pacific Rim, there is a new weapon which requires two persons to operate. The weapon is a giant robot designed to combat the bad aliens that are wreaking havoc on earth. The robot is so complicated that the processing power of one brain alone is not enough; two brains are required. And in order that these two brains can work together perfectly, a system of "mind melting" is introduced to synchronize the processing of the two brains so that they always closely collaborate in manipulating the robot as if they were one brain. In the movie two persons are situated inside the giant robot and the streams of the content of their two brains are connected and synchronized. The result is that one person can have access to the content of the other, and vice versa. This is not the case of two persons totally becoming one, because there is no possibility in the story of one person in the team becoming dominant. Instead the two work together so seamlessly that they look as if they become one, and indeed the robot itself is one very big mechanism. The story here seems to show that at least it is possible for two persons to share their mind stream yet not become one and the same. The "melting" that occurs is not a real melting of the two persons, but a close merging of the two brains' processing power in order to accomplish the complex task at hand. During the control of the robot, the two persons share very largely the same psychological continuity, their minds almost merge completely. Yet they remain different. In fact we don't have to imagine a far-off scenario like the one in the movie. In sports match we can also have something close to "mind melting" in this sense such as in tennis or badminton doubles competition. During the match their minds are so focused that they virtually function as one unit; the contents of their minds then consist of nothing but the match itself, and since it's the same match it is very likely that according to the Psychological Account they become one person. But the fact that they are not seems to belie the account.

The above examples are those about two persons functioning on the same task and somehow sharing at least a portion of the other's mental stream together at one period of time. Alternatively one could also imagine a related case where two different persons share roughly the same stream at different times. Imagine that all the content of someone, S, is copied and then downloaded onto an empty brain of another person, S*, thereby giving life to the empty brain and revives S*. All the while S remains alive and well. Do we have one or two persons here? It seems more plausible to hold that there are here two different persons because, firstly, S and S* occupy different spatiotemporal positions and have different bodies, so the most that they can be is that they are type-identical rather than token-identical. But it's very difficult to make sense of an individual person being type-identical, but not token-identical, with herself. We seem to have an intuitive feeling that, even though it were possible for there to be two individuals both of whom are type-identical with each other, we nonetheless believe that those two are not one and the same individuals anyway. Our own mundane experiences tell us that this case is just a more extreme one of identical twins (though they are not exactly type-identical, after all even identical twins are not exactly or mathematically type-identical); as we find it difficult or impossible to imagine a case where two identical twins are so similar that they merge their bodies into being one and the same, these extreme twins should not be different.

The upshot of this is that the Psychological Account at most could give only a necessary condition for personal identity, but not a sufficient one. The mere fact that one has some kind of psychological continuity with a person at another time, past or future, does not by itself entail that one is thereby identical personally with those persons at the other times. It remains for me to show what else is required, and according to the externalist view that I am proposing, external factors such as birth certificates, testimonies of those who know the person well, old photos, records of the person's writings, etc. can be used as means by which one's personal identity is established. These mementos, however, cannot establish a necessary truth that one is personally identical with another person at another time. I don't think any factor can establish that much, because the whole notion of one's being the same person with another one at another time as a necessary truth is not well explicated at all. Moreover, since one's bodies and mental episodes do indeed change continuously, it is very difficult to find any factor that is responsible for there being a necessary identity of the two persons or person-stages here. However, this does not have to concern us now, nor concern the ordinary persons as they go about in their daily lives based on the mundane beliefs that such and such persons are the same as those they encountered yesterday or a year before. I don't think this point shows that there can be no way of establishing sufficiently one's personal identity with one's person-stages at other times, only that it seems to be impossible to establish this as a necessary truth.

Against this Parfit's response is that he introduces the concept of quasi-memory. One has a quasi-memory when one seems to have a kind of memory but it's essentially a spurious one because it's a "memory" that belongs to someone else and not to me. For example, if someone's memory were to be planted into my brain I would be a quasi-memory instead of a real one. This is because, according to Parfit, the fact that I have a memory of an event in the past implies that it is I myself who experienced that event in the past. Having a memory, a genuine and not a quasi one, implies that the memory belongs to me and that I was the one who actually experienced the memorized event first hand. Against Butler and Reid, Parfit contends that it does not actually matter that I might have a number of quasi-memories. Even if I have a number of quasi-memories, or someone else's memories in my brain, the fact that it is I who does the remembering of these memories, quasi or not, is enough to show that I have personal identity through them, since the memories, quasi and all, form my psychological continuity. All this is possible because Parfit believes that having a memory intuitively implies that I am related numerically to the person in the past. According to him, "Overlapping strands of strong connectedness provide *continuity of quasi-memory*. Revising Locke, we claim that the unity of each person's life is in part created by this continuity. We are not now appealing to a concept that presupposes personal identity" (Parfit 1984, Ch. 11 Sect. 80). Thus simply having a bunch of quasi-memories is enough for me to maintain that I am the same person as the content of those quasi-memories indicates. However, that is only possible if one insists in believing that it is one's strand of memory, the sheer fact that one is having these strands of memory, that is sufficient for one to maintain one's identity. What is missing in this account is how it is possible for one to maintain it in this way. The only way possible seems to be that memory or psychological continuity is the sufficient condition, but that needs to be argued for and cannot simply be assumed. The reason why memory or psychological continuity is a powerful means of maintaining identity is because its content is vivid enough to show the subject that indeed she is identical with the one being memorized. But Parfit's use of quasi-memories here seems to belittle this important role of content of memory altogether.

3.2.2 Against the Bodily Account

As for the Bodily Account, perhaps its most sophisticated treatment of is offered by Eric Olson. In his *The Human Animal,* Olson has the following to say:

> What it takes for us to persist through time is what I have called *biological continuity*: one survives just in case one's purely animal functions—metabolism, the capacity to breathe and circulate one's blood, and the like—continue. I would put biology in place of psychology, and one's biological life in place of one's mind, in determining what it takes for us to persist: a biological approach to personal identity (Olson 1999, pp. 16–17).

Olson then elaborates:

> The Biological Approach makes two claims. First, you and I are animals: members of the
> species *Homo sapiens,* to be precise. I do not claim that all people are human animals, or
> living organisms of any other species. For all I know there are or could be intelligent
> Martians, gods, angels, demons, trolls, or even rational, conscious electronic computers
> made of metal and silicon. But all human people are animals. We are what Locke called
> "men." ... The second claim is more controversial: Psychological continuity is neither nec-
> essary nor sufficient for a human animal to persist through time. I believe that this is evident
> when we think carefully about what it means to be a living organism (Olson 1999, p. 17).

In this way Olson attempts to turn the whole problem of personal identity upside
down. Instead of looking for a set of criteria by which one can ascertain the identity
through time of a person, Olson subverts the attempt and claims that there is nothing
to *personal* identity beyond identity of a biological organism. As the latter kind of
identity is constituted by biological functions such as breathing, metabolism, and
the like, these functions are sufficient in determining the identity of a person. Olson
says: "In a sense, ..., there is no such thing as personal identity, any more than there
is such a thing as infant identity or philosopher identity" (Olson 1999, p. 27). One
may ask what are the conditions by which an infant or a philosopher continues as
infant or philosopher through time, but for Olson there are no such conditions. In the
same vein, then, to ask for conditions that would be necessary and sufficient for
personal identity beyond the mere physical presence of the organism as a biological
entity would futile too.

There are in fact many similarities between Olson's account here and the externalist one that I am proposing. First of all both do not accept the Psychological
Account; furthermore both do not accept the prevailing view, implicit in the
Psychological Account and other related ones, that there is something to identity
such that any talk of an entity persisting through time has to presuppose some kind
of existence of some entity that does the persisting. For the Psychological Account
this of course is psychological continuity; it is this thread that ties the disparate
episodes together that is responsible for maintaining personal identity. But for
Olson, rightly I think, there is nothing over and above mundane identity of a biological organism. That is, an organism is ordinarily an entity that persists through
time; it is born and persists for a while, and then it dies. While it persists it has to
maintain its viability as a living organism; it has to eat, breathe, pass out waste, and
so on. For Olson it is these activities, taken together as functions of a single organism, that comprise its identity; there is nothing over and above these functions when
it comes to its own identity.

However, it is the idea that there is nothing over and above the ordinary identity
of biological organisms that lands Olson in the most difficulty as an account of *personal* identity. We normally do not regard animals such as cats and dogs as persons,
though they are undoubtedly biological organisms. For Olson an account of the
identity of a dog or a cat would necessarily be the same as that for a human being,
indeed for a human person, since for Olson there is no difference between a human
person and a human animal. However, one can readily conceive that identity conditions for the two do not need to be one and the same. Persons can own property,
enter into contractual obligations, vote in elections, and so on, while the human

animal, qua animal, cannot. Thus one can question whether Olson is answering the same question or addressing the same problem when he claims that his Biological Approach is a solution to the problem of personal identity. Certainly everyone knows that a human person, materially, is a species of animal. But to claim that there is nothing to the identity of a human being qua a person beyond identity of the same qua biological organism seems to be too far-fetched. But if this is the case, then Olson's idea that identity conditions for biological organisms suffice as conditions for personal identity are too broad. Olson's conditions could at most be necessary (if it is necessary that a person inhabits some kind of biological body), but they are no more than that.

My argument above rests on the proposal that identity conditions for a person are not the same as ones for an animal. Here we come up against methodological issues. Olson seems to believe that thought experiments of the kind Parfit is talking about are not legitimate because they cannot be actually performed using today's technology. However in philosophy thought experiments are necessary when the problem being discussed concerned possibility rather than what is merely actual. Two things are not one and the same if it is only possible that one can have certain properties that the other cannot. Following Quine's example decades ago, animals having kidneys are not the same as animals having livers, even though in this world all animals having one organ are always co-extensive with animals having the other. But since kidneys and livers are not the same organs it cannot be said that animals having kidneys are absolutely identical with animals having a liver. There are possible worlds in which they are not the same. Olson does not accept this because he seems to talk only at the level of the actual world, where every animal that has a liver always has kidneys. Hence for human persons and human animals the situation is the same. In this actual world persons and the animals are the same (as of now), but it is possible that they are not (if thinkers such as Ray Kurzweil is right, then the moment, known as the Singularity, is near – Kurzweil 2005). But that it is possible that they are not the same is sufficient for their identity conditions not to be one and the same.

3.2.3   Against the Narrative Account

Back to the Narrative Account, Marya Schechtman argues that persons create their own identities by constructing narratives that tell stories about their lives (Schechtman 2007, p. 93). Here identity is viewed rather differently from what is being considered here. What we are concerned with so far has been the problem of how to account for the identity of a person through time (and to a lesser extent, at a time); this is more a philosophical question, a conceptual one. However, what concerns Schechtman in her book is the problem of how persons construct their own identities. In the first case, identity comes before the person, so to speak. The identity is already there, and the problem is how to account for it. In Schechtman's case, on the other hand, the identity is not already assumed, but the problem is shifted to that of constructing identities through narratives, implying that identities come after the persons themselves. We are not directly concerned with this latter notion of identity, though it features to a certain extent in our treatment of the online selves too. The point is that if Schechtman's main point in dealing with personal identity is the view that persons construct their identities, then this point is not directly related to what is being discussed here.

Schechtman has the following to say about her own view:

> According to the narrative self-constitution view, the difference between persons and other individuals (I use the word "individuals" to refer to any sentient creature.) lie in how they organize their experience, and hence their lives. At the core of this view is the assertion that individuals constitute themselves as persons by coming to think of themselves as persisting subjects who have had experience in the past and will continue to have experience in the future, taking certain experiences as theirs. Some, but not all, individuals weave stories of their lives, and it is their doing so which make them persons. On this view a person's *identity* (in the sense at issue in the characterization question) is constituted by the content of her self-narrative, and the traits, actions and experiences included in it are, by virtue of that inclusion, hers (Schechtman 2007, p. 94).

As we have seen before, creating one's own identity through stories we are making of ourselves does not logically solve the problem of personal identity as raised by Locke, Parfit and others because there is simply no criterion of correctness that would guarantee that the stories we are making are really the ones that happen to each of us. One simply goes back to Reid's and Butler's circularity objection of the Memory Account—in order for me to be certain that the story I am making of myself is one that shows that I am one and the same person throughout the story here, I need already to be one and the same person from the beginning. However, it is to Schechtman's credit that she recognizes the need for an externalist conception when she argues that it is the view of us that others make when they perceive us that is indispensable in any account of our identities (Schechtman 2007, pp. 94–96). Schechtman argues that one of the necessary constraints of the story account to be able to constitute a self is that the views that other people have of us. I might be making a story that narrate my life in order to make sense of who I really am and that I am essentially one and the same person throughout the story, but it is the views of other people, their perception of who I am and what roles I play in the community (as a teacher, father, husband, etc.) that provides a constraint that prevents me from making up any kind of stories I would perhaps like. Thus I cannot, on pains of contradicting the views that others have of me, construct any story even though it may be fully coherent. Usually one relies on one's memory of one's life experience as it runs through the course of one's life as being remembered when one constructs a biography that would fit Schechtman's view of narrative self-constitution. But we have seen that there are problems with relying on memory, and hence we need testimonies of others and other evidence and external factors to help bolster our memories. The idea is that reliability of these external factors cannot come from the subject's memory—that would certainly be circular. Instead the reliability or justification of the external factors has to be based on independent and reliable sources, such as certified birth certificates and testimonies of reliable witnesses and so on.

3.3 Online Personal Identity and the Extended Self View

We can also find accounts of personal identity in the online world. The problem here is how to account for identity of an online person across time. An analog of the memory account of personal identity in the online world would be that the criterion used to verify that what I posted online a year or two ago really does belong to me and not somebody else is that I remember doing so. I can look up what I did post 12 months ago (with the current "timeline" feature now on Facebook, it has become rather feasible to do that now, even though I post so much material on Facebook). The reason why I know that these postings belong to me, or to my online self, is because I remember doing so. When I see the postings in the past, according to the memory account what tie these postings to the sense of being myself now is that I remember posting these things some time ago, and that I recognize these postings and do remember the occasion that prompted me to post them at that time.

However, the circularity argument against the memory account discussed earlier of the offline personal identity applies in the online world too. If I chance upon a set of old photos taken, say, 12 years ago, which contain scenes of what I remember to take part in, what seems to tie these photos to me is that I do remember the scenes when the photos were taken. The photos would bring me back, so to speak, to the scene in the past and I then remember old friends that I was close with at that time but seldom see now. My personal identity from the time when the photos were taken and the time I am watching them now would be supported by the thread of memory. But perhaps there is a circularity here. People usually say that old photos "bring back memories," but how could the photos bring back memories when our memories are being used to identify the scenes in the photos as belonging to us (and not to somebody else altogether) from the beginning. We have an inkling of the circularity argument here. What happens seems to be that the photos occasion me to relive scenes from my past experiences, which I would not have been attending to were it not for the photos themselves. On the one hand, I recognize the photos as belonging to me—they are my photos; on the other the photos bring back scenes that I thought I might have forgotten all along. The memory that I use to identify the photos as belonging to me must be there and it is strengthened and given more detail and content when I examine the photos. If it were possible that my memory could be provided more detail or content in this sense, then it would appear that memory is supported by the identification that I make with the content of the photos. This does not seem to be predicted by the memory account. Furthermore, there could conceivable be cases where, as in Reid's example of the forgetful old general, I fail to remember that it was I myself in the picture. Perhaps I suffer from a bout of mild amnesia. But it is also conceivable that my amnesia is not too strong and the memory can be brought back through some means, such as when there are writings at the back of the photo describing the scene and the occasion when the photo was taken. The description, then, provides an external help to aid my memory, something that strengthens my memory. The point is that memory alone might not be adequate; the description functions as an external factor that provides more reliability to the memory. And if we substitute these old photos for old online photos and postings one has on Facebook, the general contour of the argument would still be the same.

If the Psychological Account does not seem to be sufficient in bringing about online personal identity, the Bodily Account does not fare so well either. Even though the popular view that every cell in the human body gets replaced every 7 years seems to be wrong (nerve cells in the brain, for example, never get replaced), there are enough number of cells in the body that do get replaced that the human body actually resembles an event rather than a static entity. But if the body is more like an event, then the criterion for its identity and continuity needs to be different from that of a static thing. For example, an event has to have a clear beginning and ending, which in the case of the body, of course, refers to the birth and death of the body and also of the person. Then we have familiar means by which we individuate the body, such as giving it a special status of a human person, giving it a name, a place in the society and community, and so on. These are the means by which the identity of the body and of the person is fixed.

The Buddhist account discussed the previous chapter illustrates this point quite well. Remember that for the Buddhist identity at a time is created when an amalgamation of factors is such that it accords with how that amalgamation is designated through a concept. Thus, an amalgamation of wood and metal of various shapes and sizes becomes a chariot when the elements are composed in such and such a way, in particular in such a way that the result becomes an instance of the concept 'chariot.' Identity of the chariot is not fixed through any existing element that is necessary for there to be a chariot—there is no necessary and essential element such that the chariot would not be what it is if it were missing. A chariot can lack one or more of its parts, e.g., it may have a spoke in a wheel missing, or some other parts missing, but it would still be recognized as a chariot if the remaining amalgamation fits what is commonly recognized as a chariot. However, if the whole thing is burned so thoroughly that it becomes a heap of ashes, then it is not a chariot any longer. The key point in the Buddhist position is that there is no essential element, what Aristotle calls the what-it-is-to-be, that serves as a key element in making a thing what it is. For the Buddhist what makes a thing to be what it is, what makes a chariot a chariot, for example, is not there within the chariot, so to speak, but outside as onlookers perceive the thing and judge according to their use of language whether it is a chariot or not. As Nāgasena uses the example of the chariot to refer to the problem of the self and personal identity as we have seen, the self owes its identity through conceptual use and designation by the community, there is no such thing as a necessary element within an amalgamation of physical parts and mental episodes such that the amalgamation a self or a person. This is the basic tenet of the externalist position.

The line of argument being made here is akin to the externalism/internalism debate in epistemology. In their attempts to locate the source of justification of belief, epistemologists have traditionally tried to look at the subject's beliefs, i.e., what lies internal to the subject's own cognitive field, as the source of justification. Thus we find Descartes locating the ultimate source of justification of his belief in the cogito statement through the fact, evident to himself, that it is clear and distinct to him that he thinks and he exists. However, recently many epistemologists have started to look at external sources for the justification. For example, Alvin Goldman has argued for a kind of social epistemology where the source of justification of belief is located outside of an individual and among the social interaction that the individual has with her social environment (Goldman 1999). Perhaps in the same way, personal identity has traditionally been associated with internalism—factors thought to be responsible for fixing the identity have come from internal sources such as the subject's own beliefs and memory episodes. However, one could follow the lead of the social epistemologists and other externalists in epistemology and start to argue that external factors are really the ones that fix the identity (see, for example, Goldman 1986, 1999; Fuller 1988; Longino 1990). For example, instead of trying to find the source of the identity internally, one could broaden out and try to locate the source instead outside of the subject's cognitive domain. A candidate could well be what others think of the subject in question, what their collective behaviors are like such that these behaviors taken together succeed in fixing the identity of the subject. Suppose I am not absolutely certain if the picture of a young one-year-old that I am holding is that of myself, I can certainly ask my mother. My mother's testimony (usually mothers are very good at recognizing her young child even though decades have passed) will then fix the identity of the boy in the old picture and my own self today. Note that my mother's testimony does not function merely as an aid to my memory, but it contributes directly to my identity. This is an important point for the externalism/internalism debate. In epistemology, the externalist position holds that beliefs are justified by their external relation to the environment. For example, in Goldman's view beliefs are justified by their "truth tracking" ability (Goldman 1999). The point is that the truth tracking feature of a belief does not depend on what I believe to be so; in fact what lies inside my belief or my thoughts is not relevant to the question whether the belief in question does in fact track the truth or not. Truth tracking is a feature of the world, so to speak, and remains what it is no matter whether I believe it or not. In our terms here, the role of my mother's testimony, as a feature of the world, functions as a factor partly constituting my identity.

Other clues are also possible; perhaps the picture is associated with some notes or documents that could relate back to me. These notes and documents thus serve as the external factors too. In fact these are the standard methods used by societies to identify persons in real life, such as in solving identity disputes in courtrooms. Here society seems to trust these testimonies than the mere internal reports from the memory of the subjects themselves.

Things do not need to be radically different in the online world. We could metaphorically regard the moment when someone registers her profile onto sites such as Facebook and Twitter and becomes known to the circle of people who are already on these social media as the moment when that person is "born." In the same vein, the moment when someone removes his or her profile from Facebook, thereby ceasing to engage in any activities that are performed by Facebook users could be regarded as their "deaths." All the activities during these two boundary marks represent those performed by the subject when she is "alive." Since it is very difficult in the online world to locate where the subjective domain is which is necessary for fixing identities according to the internalist theory, external factors are thus the only ones available for fixing them. In Facebook, for example, there are guidelines that one needs to follow in order to be "born" or to "die" there. One has to follow certain rules in order to have one's profile picture show up; one has to register oneself, answer a number of questions, invite friends, and so on. The "birth" of a new user on Facebook can be announced publicly throughout the Facebook world, or it can be a rather quiet birth where the subject comes on the scene quietly without much fanfare. In the same vein, Facebook also has a clear policy regarding the "death" of its user. Formerly it was very difficult, if not entirely possible, to delete someone's profile from Facebook, but after much protest Facebook then allows someone to delete his or her profile rather completely. Furthermore, it also enables users to "memorialize" a deceased user. An account that has been memorialized will remain, and the user's close friends can have access to the wall of the account to post their remembrances. Thus, in effect the wall of the deceased and memorialized user becomes a grave where close friends can drop by and pay their respect (See http://www.facebook. com/help/?faq=13016 and http://www.facebook.com/help/?faq=13941). Here, then, the identity of the person on Facebook is constituted through the information that is posed by the person herself as well as what others post about her. These are the activities that take place after the moment when the user is "born" and before she "dies" or removes herself completely from the site. Furthermore, even if she really dies in real life, her posts and comments can still be available, in the same way as the thoughts, ideas and writings of dead persons can be available to us. The postings and comments of the dead person will remain there and there will be no new additions, in the same way as a dead person cannot write a new book.

Focusing on the self, J. David Velleman argues that the word 'self' does not denote a single entity but "rather expresses a reflexive guise under which parts or aspects of a person are presented to his own mind" (Velleman 2006, p. 1). The self functions as a framework by which one can become self reflexive, representing aspects or things to oneself. Hence it seems clear that the self is not made up of the body (there is no single part of the body that is directly responsible for the self; see Klein et al. 2002), and that it is difficult to maintain that there is one particular set of internal mental episodes that is directly responsible for the self either. These episodes are just what they are, and without any means to collect them, or to bind them together in a single whole, they are just disparate episodes. On the other hand, it does appear, too, that the self cannot be other than these physical and mental episodes, even though we may not be able to find one particular physical substance or one mental episode that is absolutely identical with it. For if the self were other than the physical or mental episodes belonging to a person, it would be entirely baffling how that is possible. That would mean that our persons or our selves belong to some kind of a soul that comes to us from somewhere and animates us, making us a human being we are. This theory, however, has been discredited long time ago and runs counter to the modern scientific mindset. So unless we find a compelling reason to accept these souls, we had better put them aside.

Leaving out the soul leaves us only with either mental or bodily factors that could conceivably constitute a self. No one particular mental episode or physical event belonging to our bodies can be one and the same as the self; this, however, does not mean either that the self can exist apart from these episodes either. Hence a consequence is that it is all these episodes, taken up and bound together in one way or another, that constitute the self. The crucial point is that, since so far we have not been successful in singling out either mental or bodily elements that are responsible for the self, we have then to look outward and search for the fixing factor in the way these bodily and mental elements are assembled (Nāgasena's point from the previous chapter) from outside. Hence the fixing factor, what functions as Kant's 'I think' which accompanies all his mental representations that we have already seen, cannot coherently be one of the factors to be assembled in the first place, on pain of circularity or infinite regress. The factor has to be supplied from external source, such as written records, testimonies of mothers, birth certificates, and so on. In the online world, the counterpart to this are, of course, written records, comments, postings, photos, and so on, which together serve as something constitutive of the identity of the person in question. These, furthermore, have to be supplied with recognition as the person by the community (a point well noted by Schechtman), for the written records (or other evidence) as such would do nothing to confirm the identity by themselves without their being recognized as such by others.

3.3.1   Objections and Replies

Perhaps one objection that could be raised against the externalist view I am proposing would be that it is counterintuitive. Presumably the basic view we have when we think of the question whether my self yesterday is the same with my self as of now would be that I can remember what I did and what I thought (some of them anyway) quite easily. Or I can construct a story that tells what I did from yesterday up until the present moment. Here no external factors seem to be needed. One might object that, were a person happen to be alone, for example if he were stranded on a desert island with no one else in sight, then where could he find any external means by which he can ascertain his identity? But being alone on a desert island does not mean that he does not live among things at all. He could leave marks on a coconut tree telling himself how much time has passed, and these marks do function as an identity marker for himself. In fact any trace left by him as he makes his presence on the island serves as reminders for him that he himself existed in the immediate past and extends through time up until the present moment. Certainly he could have used his memory alone, but memory becomes then only one factor among many; as normal aids to memory becomes extensions of the same memory, his memory could be said to extend, so to speak, outward from his brain to the coconut trees and whatever else that he uses to remind himself.

The objection might go further. Suppose that the person is not stranded on a desert island, but is utterly alone. Perhaps he is suspended in outer space, far away from anything that could be used as a reminder for his memory or as any external identity markers for that matter. In this case it is hardly conceivable that the person in question would remain a viable person for long. The reason is not only the physical one of not being able to survive in outer space for an extended period of time without proper life support, but the more philosophical reason is that a person is not exactly who he or she is without relations to other persons or without her place in any community. Without a community, the person, according to Aristotle and Hegel, just ceases to be a person. One reason for this is certainly physical. Life support system that would enable anyone to survive in outer space even for a few seconds has to be researched, manufactured, procured and made available to the person in question. All these require a community. Nevertheless, if we suppose the person to survive, through some magic, in outer space for a time without any direct linking to the community or to other things, in this case what make him or her the same person throughout would be his bodily constitution and his mental continuity. But these factors would not mean anything much because a person suspended in outer space without any relation to anything at all outside of him—with no one to talk to, nothing to latch on to, and so on—would be in a very dire predicament. And if Aristotle and Hegel are right that a person derives her meaningfulness from her relations to others, then the person's own identity would then not seem to matter much at all in the case of being suspended in this way. Furthermore, all the caveats that we have seen earlier regarding the use of memory and bodily constitution do still apply in this case too. Memory can fail; the body does deteriorate fast in outer space (which is why we need a large dose of magic to make this thought example work)—these show that these two factors do not work in themselves. Moreover, we seem to forget in our discussion here that our suspended person does still have an external relation after all. It is we, the readers of my text here who are now thinking of the suspended person, who provides such relation to him. This does not seem to be as frivolous as it might first appear. The person who has nothing at all to relate to would in principle be unconceivable, because in order to conceive of him or her at all, he or she needs to have at least a connection, which *ex hypothesi* is not possible.

This shows that personal identity is a pragmatic concept and does admit of degrees. That is, one can coherently talk about a person's being more or less identical with his former or future selves. If the evidence is strong, then she becomes more identical, or more identifiable, with her self in other times, but if it is weaker, then the identification obviously becomes weaker too. This point follows from the Buddhist position alluded to earlier. When the self or the person is itself a construction out of various elements and episodes, then whether this self or that person is identical with her former self would be a matter of how much these elements and episodes are retained between the two selves or two self-stages. Language purists and logicians certainly would balk at the talk of something being more or less identical with itself or with another thing. For them a thing is *either* identical *or* unidentical with another, no degrees allowed. In order to not to quarrel with them, we use the term 'identifiable' instead. A person can be more or less identifiable with her former or future self just in case there are more or fewer factors that make her up to be the person that she is at one moment that are retained through time in those former or future selves. The more retained factors, the more identifiable the two persons become, and *vice versa*. And since factors that make up a person do also come from her relations to other things, then markers of her personal identity do come from these factors outside of her skin and her brain too. For illustration, the testimony of the person's mother that the person is indeed her first-born child, if consistent for person-stage1 and person-stage2, would then confirm that the two person-stages are in fact one and the same person. If we have more of these, then the case for the two person-stages to be one and the same person become stronger — these factors can come from within the memory field or outside of the brain or the skin, as we have seen.

3.4 The Informational Self and the Role of the Body

In *Remembrance of Things Past,* Proust has the following to say:

> But then, even in the most insignificant details of our daily life, none of us can be said to constitute a material whole, which is identical for everyone, and need only be turned up like a page in an account-book or the record of a will; our social personality is created by the thoughts of other people. Even the simple act which we describe as "seeing some one we know" is, to some extent, an intellectual process. We pack the physical outline of the creature we see with all the ideas we have already formed about him, and in the complete picture of him which we compose in our minds those ideas have certainly the principal place. In the end they come to fill out so completely the curve of his checks, to follow so exactly the line of his nose, they blend so harmoniously in the sound of his voice that these seem to be no more than a transparent envelope, so that each time we see the face or hear the voice it is our own ideas of him which we recognize and to which we listen (Quoted in Floridi 2005, pp. 194–195).

This passage neatly sums up the basic idea of the self as constituted through information. The "material whole" of our bodies is not everything that matters. Our personalities and things that make up who we are only partially constituted by our bodies (we may be characterized by our peer as being skinny, plump, tall, fair, dark, and so on) and the rest are by the images that others have in their minds. We present a *persona*, a mask that we always wear, intentionally or not, to the outside world; that cannot be separated from any account of the self. "We pack the physical outline of the creature we see with all the ideas we have already formed about him" — the idea we have of others consist of beliefs that we have about that particular person. For someone we already know, we already have a number of beliefs about that person, and each time we encounter that person, the set of beliefs will largely be confirmed, or revised piecemeal when some aspects of her personality do not match our expectations. It is very rare for some person to show up with none of the prior beliefs about her confirmed; perhaps this is not possible as a psychological fact. But all this shows that our persona are constituted by information. The beliefs and preconceptions we have of someone we know are information about her, and as we have shown in the previous section, it is the information that we have of others that play a necessary role in constituting who she is. For Floridi this is the core idea that the self is constituted through information (Floridi 2011b, pp. 7–9). And I believe that Floridi is largely right on this.

In the Buddhist thought that we have been discussing, the self is a construction in that it does not have inherent existence. In the last chapter we reviewed the *Anatta Lakkhana Sutta* and found that whatever is to be the self lacks the necessary property that would enable it to be such, that is, the ability to be controlled according to the will. Our bodies are not subject to our will, and more interestingly our minds are not subject either because we normally cannot keep up our concentration for any sustained period of time, even if we want to do so. The 'we' that function in the clause "we normally cannot keep up our concentration" is according to Buddhist philosophy an illusion because the language gives an appearance of there being a subject of the sentence, one who performs the functions referred to in the predicate, or in this case one who does not have the ability to perform sustained concentration. However, this is also an illusion because when we reflect on the subject we find that it is also subject to the same analysis. It might be posited that there is another subject who performs that looking back to the original subject, but that would invite a vicious infinite regress. Suffice it to say, then, that the self is a necessary component of thought, for without the ability to form a proposition, which necessarily consists of the subject and the predicate, no thought would be possible. But as we have seen the subject's role in a proposition does not imply that there is a self which inherently exists as a substantive entity in its own right. It functions more as a place holder, so to speak, in order that the grammatical construction of the proposition be possible. Hence, the self in this sense is constituted through information because it is part of the information that is contained in the proposition in its role as the subject. The identity of the self here is numerical, for the question is how a self, as subject, is constituted as something identical to itself at a time. And we have seen that in Buddhist philosophy numerical identity of the self is from the beginning a fiction because it necessarily consists of various parts, each of which has their own separate identity and it is only when these parts are assembled in a particular way that the self emerges, in the same as the various parts of the chariot, which when assembled correctly become parts of a functioning vehicle. We could also see that the particular way the chariot parts are assembled represent a particular piece of information, one that enables the chariot to be assembled. One is reminded as a manual for building up a chariot—one has to put this part there and that part here, and so on. The manual obviously contains information, and that information is reflected in the way the chariot is correctly assembled.

That the self is constituted through information does not imply that the self is something ethereal and abstract; on the contrary it can be as concrete as it can be. We might compare this with the recent technology of 3D printing. It is now possible, as it perhaps quite well known now, to "print" three-dimensional stuff in the same way as we traditionally print stuff on two-dimensional pieces of paper. The product will become a three-dimensional object whose specifications and shapes are determined by the software on the computer that instructs the "printer" to produce the 3D object. I put the word "print" in quotation marks because normally we do not conceive of printing as producing some three-dimensional object. (However, this parlance has entered into the common discourse in English so that it does not sound strange at least among those in the technical world to print out some three-dimensional object.) The idea is that a 3D object certainly has a body, but that body is entirely constituted through information. Every aspect of the object is determined by the software on the computer, which as we know is an entirely information manipulating machine. This seems to show that the body itself, whose essential property of course is extension, can be constituted through information. Thus if the body is part of the self, then it can be constituted through information too and the body then is part of the informational self. Moreover, when we examine the human body, we find that it is constituted through information too. The cells that comprise the body are produced according to instructions given in the DNA, a long string of codes that contain an enormous amount of information.

However, the objector might say, counter to this, that it may be true that the body is built upon specifications or information given in the DNA. That is one thing, but it is another thing all together to say that the self in and of itself is information. What is objectionable, so the argument goes, is the latter view rather the former which after all is an obvious scientific fact. Nevertheless, as we have seen from the previous discussion of the Buddhist view, the self does not exist as an inherently subsisting and enduring entity. There is no pit that lies inside when we take away all the flesh, for example. The mind-body complex is more like an onion, which yields nothing inside beyond the concentric layers of the same. That, on the contrary, does not mean that there is no self, more precisely no self that we always refer to whenever we use the first-person pronoun. Hence the self itself is informational, according to the Buddhist stance; it is not merely something constructed out of a set of instructions or information, though it can readily be regarded as such. As it functions as a unifier of experiences or a fulcrum point that ties the narrative together to form a coherent whole. Dan Dennett believes that the self functions as the abstract center of gravity point in an object (Dennett 2013, pp. 333–340). Every object has its center of gravity, a point where the object balances itself as the forces pulling on it do so equally on all sides. The point cannot be seen nor touched, but it is already there because we can calculate for any object where its center of gravity is. It is there when external forces acting on the body are considered and the point where these cancel each other out are calculated. For Dennett, the self acts in the same way; it cannot be perceived in any way, but it is there as a point where all the forces cancel each other out and in the case of the self where the various episodes of the mind-body complex are unified into a coherent whole (Dennett 2013, pp. 333–340). But if it cannot be seen or weighed or tough, then it is quite clear that the self is even informational in this way.

However much Dennett's view on the nature of the self commends itself to the Buddhist position in that the self is not posited as a really existing entity, the main difference between his view and that of the Buddhist is that in Buddhism the self cannot be located, whereas in Dennett's view it certainly can, as center of gravity can be located of any object. The Buddhist does not deny that the brain is responsible for all the thinking of the human being, but this does not imply that the self must be located in or at the brain. Persons who suffer from total amnesia do not remember who they are, and it is possible that they can assume an identity of a new person even though their bodily constitution does not suffer any major disruption.

Their bodies might not suffer from any injuries and could well carry on as a normal human body always does, the difference being that before the amnesia she was one person and after she was, practically speaking, another. That she has become another person is normally judged by her social peers, who accept her into their fold with a view of her being this particular person and not any other. It is entirely conceivable that before the amnesia she had another group of friends who remember who she was before the amnesia but as they realize the amnesia is irretrievable they accept that the old person is lost and the new person who emerges after the amnesia is, practically speaking, another person who interacts with another social group. The person herself might even recognize this, but she regards who she was before the amnesia as totally another person, even though the bodily constitution remains the same. If this story is plausible, then the brain itself, as a physical organ, is not responsible for her being one particular person and not another, for her brain, as a physical organ, does carry over from the pre- to the post-amnesia stage. Her brain was not transplanted with another one that belongs to another person; she just assumes the identity of a new person. This does not imply that the brain has nothing to do with her personality or her memory loss, far from it, but it does imply that the brain is not sufficient, and her recognition by her social peer, her role within the community, in short the socio-cultural context surrounding her person is also responsible. Since that context cannot be physically located or pinpointed, the self or her person, as partly constituted by her recognition and her role in the community, cannot be as readily pinpointed as can the center of gravity. We can also view this situation as that of the social context determining or recognizing who the person is (of course in the eyes of her friends and acquaintances, as is the case in our normal lives), as that of the social group providing and supplying information about her so that she has a standing within the social perspective of her group. As Proust says, one has a set of images that one always has of another in order to form an overall picture of who the latter is, these images are constituted by information—who the person is like, and so on, and as the self of the person lies in the recognition of her peer (a point very well made by Hegel in the Phenomenology of Spirit (1977)), then we can conclude that the self can also be informational in this way.

Viewing the self as made up of information makes it easier to account for the self in the online world. Someone enters the online world, setting up an account on Facebook, uploading her portrait, signing up with her email address, invent a password, receives a number of "friend" suggestions, and send friend requests to some of them. All this establishes her online presence; in other words, setting up her "self" in the world of the social network. In the later chapters we will see how this can be manipulated, and people do the manipulation of their online presence in very creative ways. Here, however, we are focused on the metaphysical problem of the status of the online identity, especially the idea that the self is constituted by information. A problem that needs to be discussed here is whether the online presence is continuous with the original self who put up the Facebook profile, or whether the online self can be its own identity which can be considered separately from the original one. This topic will be taken up in details in the chapters following this one.

3.5 Externalist Theory of Personal Identity and the Extended Self View

The externalist view that I am proposing is essentially an externalist theory of personal identity. That is, I am proposing that factors that constitute an identity of a person either through or across time are external ones. These factors are external to those which are normally considered to be constituents of the self or the body, namely the physical body itself and the subjective mental episodes of that person. Since the body does change and does not have to contain the very same elements or particles that make up the physical stuff all the time, the body alone does not appear to be the candidate. On the other hand, mental episodes are notorious for their changeability and I have tried to show that Locke's view that identifies the identifying factor with memory does not work. What seems to work, in my view, are a collection of external factors that people actually use to identify themselves and their counterparts in real life, such as testimonies of friends, documents such as birth certificates, old photos, recognition by those who know the person, and so on. This view, furthermore, also bears some relations to the view, proposed by Andy Clark and David Chalmers that the mind is extended (Clark and Chalmers 1998). According to Clark and Chalmers, the mind does not merely exist inside someone's brain, so to speak, but as we extended our mental capabilities beyond what is naturally available to us (through the use of written records and smart phones, for example), there is a real sense in which our minds are extended to these devices also. The difference is that my view, what I call the Extended Self View, is focused on the problem of personal identity, while Clark's and Chalmers' is one on personal ontology. The difference between the two is that the problem of personal identity seeks to understand how a person remains the same through time, to the extent that this is possible. Personal ontology, on the other hand, is a problem of how to explain the ontological status of a person. The two problems are thus related, but not the same.

However, I am also proposing a view in personal ontology, the view that the self is informational. This view has a lot of affinities with externalism regarding personal identity as well as the extended mind thesis. As we have seen, there appear to be strong supporting reasons in favor of the view that the self is informational in nature, and as the information does not have to be restricted to what is inside the body or the mind of the person, then the information that could constitute the person can come from outside of her own skin too, which is very similar to what the extended mind thesis says. Let us look, then, more closely at Clark's and Chalmers's view.

In the old days, people relied on chalk and tablets made of polished stone as an extension of the mind when they do calculations, and today they use pocket calculators or their computer tablets to do the same thing. In both cases Clark and Chalmers see that the mind is extended outward because either the stone tablet or the plastic tablet functions as an extension, a heuristic device, whereby the works of the mind is aided. Furthermore, the functions of these tablets are not limited only as an aid to the mind, but the functions of the mind itself are there in these tablets too.

In these senses the tablets perform some "epistemic action" (originally from Kirsh, D. and P. Magilio (1994)) and thus deserve some credits of their own. If the brain is viewed as a giant collection of neurons and is traditionally credited as a the seat of the mind, then if rock tablets or computer tablets perform the tasks that can be viewed as an extension of the mind's work, then due to the recognition of their epistemic action some form of credit should be granted to them. As Clark and Chalmers say, "epistemic action … demands spread of *epistemic credit*" (Clark and Chalmers 1998, p. 8). Even the use of the arms and the hands, which are often employed when the brain is engaged with some tasks, could be seen as an externalization of the mind outward, and in this case the body is employed in helping, indeed in sharing, the work of the brain. If the works of the hands in concretizing and externalizing the work of the brain is an integral part of the work of solving the mental task at hand, then as epistemic credit should be given to these organs it should be fair to claim that the arms and hands are part of the mind too. Sometimes one thinks by verbalizing what one is thinking—one has in mind the image of someone's talking to oneself when being engaged with a mental task. In this case the verbalizing out loud and the thinking that goes on inside the brain are almost one and the same. And as we seem to be more readily accepting of the mouth and the tongue as an externalization the work of the mind, then we should be more accepting of the arms and the hands, and by extension devices such as the chalk, the rock tablet, or the smart phones and computer notebooks too. These are all parts of the extended mind.

Hence when one thinks about personal identity in the context of the extended mind, the problem is then how to account for identity of a person, or a self, whose mind is extended outward. If we believe, as do Clark and Chalmers, that memory does not exist only inside the brain, then we accept that the person, if identifiable by her memory, is also identifiable by these external factors. In other words, my externalist account of personal identity can be seen as an extension of the Memory Account when the location of the memory in question is not limited only to the relevant brain, but notebooks, memos, smart phones, tablets and other memory enhancing devices play a constitutive role too. If indeed the person is constituted by her memory, then as memories can spread outside to these devices, then it is these devices that constitute her memory too. These clearly are not subjectively introspectible memories in the same way as information inside the brain is for the owner of the said brain, but since memos and information inside tablets can be accessed by other people they can be more public in this way. However, as it has become more technically feasible to gain an access to the content of someone else's brain and for someone's brain to send out signals outside of the body, then the line between what is subjectively introspectible (the subject domain of thought, intention) and the objectively or publicly examined (memos and records, signals from someone's brain that spread outside of the skull) is getting blurred. This does not mean that privacy cannot be protected; in fact one can protect the privacy of one's information and memories by putting the documents inside a locked safe or protecting the information inside the computer with a password.

However, in a recent paper, Eric Olson argues that the extended mind thesis rests on a mistake (Olson 2011). For Olson, the seat of identity is firmly one's physical and animal body, as we have seen; hence he naturally views the whole idea of the extended mind thesis and presumably also the informational self view in a bad light. His argument is basically that if the mind or the self can be extended outward, then there would be two thinkers, namely one which is the biological organism that does the thinking, and the person that is separated from the organism. For Olson this predicament befalls every philosophical position that separates the person from the body or the organism. The extended mind thesis implies that things outside of one's body can be parts of her mind or her own self, and for Olson this means that the person here is not actually an organism. This is because an organism retains some properties that do not change when these external things are changed. For example, when the notebooks that contain parts of the person's memory are taken away, the biological organism does not get shrunk. Thus on the one hand, something gets shrunk when the notebooks are destroyed or taken away, for Olson this is presumably the person herself, but on the other hand, it does not mean that the organism itself does get shrunk when the notebooks are taken away. Notebooks are outside of the body, and the body is the organism, so when notebooks are seen to be parts of the person, the upshot then is that the person would not be one and the same with the body. For Olson this entails a number of telling objections, chief among which is the problem of too many thinkers. If the person here is not the same as the organism which forms her existence as a biological being, then the organism does think, and the person does think too. However, both the person and the organism are actually one self, so it appears that one self is divided into two incompatible aspects, and hence this is the contradiction that leads Olson to object to the extended mind thesis. Apart from the objection about the person or the organism is getting thinner when the notebooks are taken away, another problem brought up by Olson is that if, as the extended mind thesis says, there are some mental states of the person inside her notebook, then the organism cannot be the subject of those states, because the organism is limited to only within the confine of its skin. So what this ends up with is a situation where the person and the organism have different beliefs, implying that there are two persons, or two organisms. For Olson looking at this in either way is absurd because there is only one person or one organism to begin with.

Talking about Otto, who records some of his memories on his notebooks, Olson has the following to say:

> The extended self implies that O, the organism associated with Otto, is not a subject of the external mental states located in Otto's notebook. They are Otto's beliefs but not O's. They could not be O's because they extend beyond his boundaries, and mental-state internalism, which is a corollary of the extended self, rules out any being's having mental states extending beyond its boundaries. So although Otto believes that the museum is in Bloomsbury, O does not: at most O may believe that the museum is wherever the notebook says it is. Yet we should expect O to share Otto's 'internal' mental states, since they all lie within O's boundaries. So it looks as if Otto and O are psychologically identical apart from the beliefs in the notebook (Olson 2011, p. 487).

And Olson intends the situation in the last sentence to indicate an absurdity. The *reductio* works from the premise that Otto has been extended to his notebook, which is an implication of the extended mind or extended self thesis, which when implies the absurd consequence, would have meant that the thesis cannot be accepted. However, Olson's argument here hinges on the assumption that the organism, O in this case, cannot be the subject of the mental states inside the notebook because of the principle that parts cannot extend beyond the entity of which they are parts. But this can only be tenable if the notebook or its content is not part of the organism from the beginning. The assumption, to state it more clearly is that, since Otto is an biological organism, he does not naturally have the notebook here as a part, so any content inscribed in it is not anything that he can be subject of in the same way as he is always the subject of the thoughts inside his head. But to counteract this normal attitude is precisely what the extended mind thesis is trying to do. What the thesis tries to argue for is that one does not have to believe that the self or the mind must forever be limited inside the skin or the brain, but Olson seems to take as an assumption what the extended mind thesis aims to argue against. At least it seems that Olson needs an independent argument to show why an organism must always be limited in its being to the skin. He tries to do this by invoking what he calls the mental-state internalism, which states that a being's mental states must be located within it (Olson 2011, p. 484). But this principle says only that if some states do belong to an organism from the beginning then those states must stay within the boundary of that organism; it does not say that the content of Otto's notebook cannot be part of his own self or his person. If we conceive of a new notion of organism, one that can have notebooks as parts, then there is nothing wrong in having Otto's notebook as part of who he is. The fact that the notebook just came to be added to his being recently only corroborates the platitude that parts of our own body or our own self can be added or deleted all the time.

We might have another look at Olson's argument and see how it is untenable. Suppose that Otto's brain is physically enhanced by adding some billion more neurons into his existing network inside his brain. This addition helps him memorize much more information than he ever could have before. Are we to say then that there are now "too many thinkers"? According to Olson adding these new neurons would seem to result in there being at least two thinkers, i.e., the original Otto and the enhanced Otto. Since the principle of mental-state internalism says that mental states cannot extend beyond the being who has them, the mental states of the enhanced Otto are then not accessible to the original Otto; the two Ottos here, the original and the enhanced ones, would then be forever estranged. This appears contradictory, because after all it is one and the same brain, only that it has been added some more neurons, so it must be the same organism. Just adding these neurons should not result in there being two persons as Olson's argument seems to imply.

However, Olson might object to this by saying that adding more neurons to the brain is a very different act from putting a notebook or a smart phone as a memory aid. The first one intuitively extends the function of the brain itself, thus it should still be the same organism; the second act, on the other hand, kind of jumbles two disparate things together and tries to make them one thing, hence the two acts should not be considered in the same breath. This distinction has a lot of intuitive appeal, but if we look more closely at the matter we find that the intuitive appeal here rests on the assumption that when a thing is enhanced (such as when a brain is added more neurons or a notebook is added more sheets, for example), those more things that are added to the original thing (the neurons or the additional sheets) should somehow look the same or of the same type as the original thing. Brains are composed of neurons and notebooks are composed of paper sheets, so adding more of the same should result in the thing remaining the same kind of thing it already is. However, this assumption does not always work, and we find ourselves in occasions where adding different kinds of things does not intuitively result in there being a new kind of thing. Adding some gold leaves to a notebook should not make it a completely new thing. It will become a much more expensive, gold-laden notebook, but it is still a notebook. Even adding sheet laced with thin computer chips, perhaps blending the chips with the gold sheets, to enhance the function of the notebook would not make it another thing beside a notebook either. In the same way, adding some extra things to the brain so that the brain can extend its function or capability does not, necessarily, have to result in there being a new organism, or a new entity in the way that Olson implies. Implanting computer chips inside someone's brain to aid her brain functions should not have to result in the person becoming a new person all together, and if we can imagine, according to the Extended Mind Thesis, that there does not have to be anything different in principle between having a chip planted inside the skull and having a notebook or a smart phone that serves the same function in terms of their functionality of extending and enhancing the original capability of the brain, then there does not have to be anything fundamentally different between adding more neurons to the brain and extending its function or enhancing its capability externally in this way.

The discussion here is much linked with Olson's basic view that we have discussed earlier. According to Olson, what accounts for an identity of a person is the sheer fact of her animality, her constitution as a biological organism, and whatever does account, scientifically speaking, for the identity of a biological organism does account for *personal* identity also. But there is a well-known problem of how to account for the identity of even very primitive biological organisms such as the amoeba. The identity of the single-celled organism when it moves about in its locale is perhaps simple enough to understand, but when the amoeba divides and turns into two amoebae, then a difficult conceptual problem arises. Does this mean that the original amoeba dies and out of the dead body of the dead one emerges two completely new amoebae? Or does the original one survives and out of it emerges a new one, something like budding? Or does the original amoeba survive, as *two* different organisms, in the newly emerging organisms that resulted from the division? There is not enough space in this book to discuss this fascinating problem (Parfit discusses about it briefly in his book), but my purpose is merely to show that if there are vexing problems even at the level of cell division, problem of how to accurately account for the identity of a single-celled organism and by extension all multicellular organisms (because they are all composed of individual cells, which increase their number always by division) is not a simple one of just looking at the big organism and decides its identity. The problem here seems to show that identity of an organism is more a matter of our own conceptual imposition on the organism rather than just a simple biological fact. It is we who regard organisms in nature as identical or unidentical to whatever, or anywhere in between. If this is so, then it is not just a biological fact for an organism to be identical with anything, it is more a conceptual and philosophical matter. And if Chalmers's and Clark's Extended Mind Thesis is tenable (which I believe it is and have argued for it throughout this chapter), then a way seems to be open to regard the notebook, for example, as a part of the person of Otto too. Basic to this view is, of course, the view that the person is not a biological concept and the person is usually broader than her physical body. If the person is not a biological concept, then Olson's account is too broad because his view does not make a real distinction between the *person* and the biological being.

3.6 The Extended Mind and the Extended Self

So in this concluding section I should state clearly what the Extended Self View is and how it is different from the view put forward by Chalmers and Clark. Essentially they are not much different at all, as my view is only an extension and a development of Chalmers and Clark. Imagine that not only the artifacts that are used by the mind such as the notebook, but the seat of consciousness itself, is extended outward, then one finds the different, if there is any, between my view, which I call the Extended Self View, as that of Chalmers and Clark. Basically speaking, if it is tenable that the seat of consciousness can be extended outside of the brain, then a case can be made toward a view where the self itself can be so extended. In fact the experiments where the subject has the feeling that they are outside of their own body, such as when the subject reports seeing their own body from an outside perspective, seems to show that the location of the seat of consciousness outside of the brain is possible (See, for example, Lewis 2015). In this case the extension that Chalmers and Clark talk about also includes that of self-consciousness; this is understandable because if the mind can be extended outside of the body, so can self-consciousness because the latter is certainly a part of the former. Furthermore, I have discussed the recent scientific research on brain-to-brain integration, where two brains are connected together as if both were nodes in a computer network (Hongladarom 2015). In this case information at the synaptic level presumably travels directly from one brain to another, creating a possible scenario where one brain is "extended" toward the other. It would be interesting to speculate on how the two selves belonging to the two original brains will feel or whether it is possible for the two selves to merge into one. I discuss the possibility of this in the context of computer games in Chap. 6. But if this were possible at all, then it is indeed possible for the self to extend outside of the body and the brain. That is Extended Self View in a nutshell.

The relevance of the Extended Self View and online self is that the latter acts more and more like avatars one has when one participates in a computer game. The richness of the information on one's profile page and the intensity of interaction of everyone on social media sites seems to make it the case that the profile itself takes an increasing role in representing the user herself, and we have seen that this is a sense where the self or the identity of the user is reflected on the self or the identity of the profile or the avatar. Thus there is a sense in which the self of the player or the user *extends* toward the self as exists on the social media. Of course this is not to say that the online self in this sense is conscious. The situation is not one where the "soul" of the user takes leaves of the user's own body and inhabit a new home in the online world, nor is it one where the online self, i.e. the profile of the person on the social media or the avatar in a game suddenly becomes conscious. That is certainly absurd. However, the sense I am making here is that, as the research on out-of-body experience and as the Extended Self View show, there is a sense in which the self does not have to be always attached to a brain inside a skull. As the avatar is capable of talking, listening, and understanding their fellow avatars in the online arena, or as a user engages with many of her friends on Facebook through her profile page and online persona, there is a sense in which the avatar and the online presence in Facebook take on some resemblance of being conscious. This situation would be more pronounced if we see that the online environment, such as the Facebook platform where everybody seems to meet nowadays, takes more and more role as a substitute or a complete reality environment all by itself.

References

Butler, J. (2008). Of personal identity. In J. Perry (Ed.), *Personal identity* (2nd ed., pp. 99–106). Berkeley: University of California Press.

Clark, A., & Chalmers, D. J. (1998). The extended mind. *Analysis, 58*, 7–19.

Dennett, D. C. (2013). *Intuition pumps and other tools for thinking*. London: Penguin.

Floridi, L. (2005). The ontological interpretation of informational privacy. *Ethics and Information Technology, 7*(4), 185–200.

Floridi, L. (2006). Four challenges for a theory of informational privacy. *Ethics and Information Technology, 8*(3), 109–119.

Floridi, L. (2011a). The informational nature of personal identity. *Minds and Machines, 21*(4), 549–566.

Floridi, L. (2011b). *The philosophy of information*. Oxford: Oxford University Press.

Fuller, S. (1988). *Social epistemology*. Bloomington: Indiana University Press.

Goldman, A. (1986). *Epistemology and cognition*. Cambridge, MA: Harvard University Press.

Goldman, A. (1999). *Knowledge in a social world*. Oxford: Oxford University Press.

Hegel, G. W. F. (1977). *Phenomenology of spirit*. (A. V. Miller, Trans.). Oxford University Press.

Hongladarom, S. (2015). Brain-to-brain integration: Metaphysical and ethical implications. *Journal of Information, Communication and Ethics in Society, 13*, 205–217.

Kirsh, D., & Magilio, P. (1994). On distinguishing epistemic from pragmatic action. *Cognitive Science, 18*, 513–549.

Klein, S. B., Rozendal, K., & Cosmides, L. (2002). A social-cognitive neuroscience analysis of the self. *Social Cognition, 20*(2), 105–135.

Kurzweil, R. (2005). *The singularity is near: When humans transcends biology*. New York: Viking.

Lewis, T. (2015). *Out-of-body experience is traced in the brain*. Retrieved from http://www.livescience.com/50683-out-of-body=illusion.html

Longino, H. (1990). *Science as social knowledge*. Princeton: Princeton University Press.

Olson, E. T. (1999). *The human animal: Personal identity without psychology*. Oxford: Oxford University Press.

Olson, E. T. (2007). *What are we: A study in personal ontology.* Oxford: Oxford University Press.

Olson, E. T. (2011). The extended self. *Minds and Machines, 21*(4), 481–495.

Parfit, D. (1971). Personal identity. *The Philosophical Review, 80*(1), 3–27.

Parfit, D. (1984). *Reasons and persons.* Oxford: Oxford University Press.

Pyle, A. (Ed.). (1999). *Key philosophers in conversation: The cogito interviews.* London: Routledge.

Reid, T. (2008). Of Mr. Locke's account of our personal identity. In J. Perry (Ed.), *Personal identity* (2nd ed., pp. 113–118). Berkeley: University of California Press.

Schechtman, M. (2007). *The constitution of selves.* Ithaca: Cornell University Press.

Schechtman, M. (2012). The story of my (second) life: Virtual worlds and narrative identity. *Philosophy of Technology, 25*, 329–343.

Velleman, J. D. (2006). *Self to self: Selected essays.* Cambridge/New York: Cambridge University Press.

Chapter 4
The Online Self and Philosophy of Technology

In the previous chapters we have seen a brief history of the self and a proposal for an externalist theory of personal identity, one where the criteria for personal identity lie in the factors external to the self or the person in question. In this chapter we will look at how philosophy of technology views the situation of the online self. Philosophy of technology, as is well known, consists of attempts to engage philosophical critique and analysis to the phenomenon of technology, itself a very complex phenomenon that can be analyzed very deeply and in various ways. The online self is saturated with technology. Not only does the computer that houses the browser and the social network sites on which the online self becomes visible and active, but the browser and the social network sites themselves are very complex pieces of software which requires million and million line of codes. Both the hardware and the software are so complex that there are tools that work on other, less abstract tools, down on to the level of the physical structure itself. There are machines that manufacture the computer chips that power the social networking sites and the browser that display them, and these machines are operated on by other machines, resulting in a wholly automated process where human involvement is only on the design stage. This trend is perhaps more visible on the software side. The codes that drive the computer chips in the personal computers today and very complicated sets of instructions, and no human being is capable of directly writing them. Hence the task is divided into modules and submodules and so on, and there are further pieces of software that help the programmer to produce the codes while the latter focuses only on the abstract task that normal human beings can understand. This total infusion of technology at many levels deserves a thorough analysis, which philosophy of technology attempts to unravel. In this regard, then, the online self can be looked at from a large variety of angles, all of which are within the purview of philosophy of technology.

We can put these angles in a number of broad groupings, viz., the metaphysical, the epistemological and the ethical. Philosophy of technology has a lot to say to all of these aspects. In fact the metaphysical side of the online self has been covered in some detail in the last chapter—the online self is a new phenomenon that demands

© Springer International Publishing Switzerland 2016
S. Hongladarom, *The Online Self*, Philosophy of Engineering and Technology 25,
DOI 10.1007/978-3-319-39075-8_4

careful philosophical analysis. However, in philosophy of technology, the metaphysical analysis of the online self is not a direct concern. What is added to the analysis we found in Chap. 3 is that a critical perspective, one that tries to analyze the online self as a *technological* phenomenon. Thus the question is not merely: "What is the metaphysical analysis of the online self?" but "As the online self is a technologically saturated product (or entity, or phenomenon, or even process), what kind of analysis can be offered from the vocabulary afforded by philosophy of technology that helps us understand its role as a piece of technology (i.e., sophisticated computing technology), or as a phenomenon or a manifestation of Technology (i.e., the global phenomenon of technology taken as a monolithic whole such as one finds in Heidegger)?" We can see that this is a hugely complex question, and this chapter can only touch upon a beginning of an attempt to unravel it.

We will start doing this by looking at how some important philosophers of technology look at the issue of online self or in case they do not have written anything directly on the topic, their views will be interpreted to see how they would have thought about the issue based on their published views. We will examine the views of Heidegger, Marcuse, Borgmann, Ihde, Dreyfus, and Feenberg, and end with a concluding section where we will try to sum up what philosophy of technology basically has to say about the phenomenon. I will also present my own view about the situatedness of the online self within the matrix of the concerns of philosophy of technology there. Basically speaking, my own view is that the online self represents neither the all menacing peril nor the unabashed promise of the Internet. That is, I do not view the online self as an expression of the total domination of Technology in our lives in the same way as Ellul or Marcuse or Heidegger appear to do; on the contrary, I do not view it as an expression of unalloyed and uncritical belief in the power of technology to bring up good life. The truth, as is very often the case, lies somewhere in between. The online self may be a manifestation of the somewhat deleterious effect of the Internet in our lives, but at the same time they can become powerful tools in shaping up and making our desired goals and our values possible. The trick is to learn how to navigate. However, before we go on fully into that discussion, let us examine the views of these philosophers, starting with Heidegger.

4.1 Martin Heidegger

A sustained discussion of philosophy of technology, or a course in the topic, cannot be complete without reading Heidegger's *A Question Concerning Technology* (1977). This seminal text is the standard of a course on philosophy of technology worldwide. In this text Heidegger lays out a powerful analysis of technology which finds echoes in many of the subsequent philosophical works on technology. For Heidegger, the term 'technology' is derived from the Greek *technē*, which means roughly art or skill in producing something. Thus one says that one has *technē* in making shoes or building houses or weaving cloth. *Technē* in the Greek mind is in contrast with physis or nature itself, without human intervention. This is an important distinction for Heidegger because the artificial nature of *technē*, the fact that what is produced through art and craft always requires human intervention and human reason, means that what is produced is always infused with the human, whereas what grows naturally does not depend on the human. Hence Heidegger sees that the product of technology, in this sense of art and craft, is inherently human and is thus a *revealing* (*alētheia*) of what is basically a human characteristic. When one weaves a piece of cloth, for example, one puts into it human elements such as the patterns and the care and other human traits which marks the cloth from natural products one finds in the wild. This act of revealing inherent in the technological product means that technology is intimately connected with truth, for truth is itself a revealing — one learns something true or one learns the trueness of something when that thing is revealed to us in its entirety without anything hidden. Thus Heidegger says that technology is itself a truth: it reveals what is distinctively human to the world.

However, this revealing aspect of technology works only for pre-modern technology for Heidegger. The technologies of the mechanical clock, the water wheel, the windmill and so on, are inherently revealing in this sense because they themselves are manifestations of art and craft where human skills are evident. These technologies are integral to the human end because they fit seamlessly well with the natural human condition. This is the condition of human beings when they are themselves parts of nature. Technology happens when humans want to achieve certain of their goals–they need clothes, hence comes the technology of weaving and producing the fabric. However, this aim of satisfying the goals does give rise to the conception of technology in the modern world where the means-end relation is everything, which led to the conception of the sovereignty of the efficient above all other values. For Heidegger, the revealing nature of pre-modern technology points to its essence, namely its *enframing* (*Ge-stell*). This is a constraint put on the work of technology so that it works in revealing its true essence to us. For pre-modern technology, its enframing would be that it reveals human nature in its wholeness and interconnectedness with nature to us. This is the truth that is revealed and whose process is the enframing.

According to Heidegger, "The manufacture and utilization of equipment, tools, and machines, the manufactured and used things themselves, and the [social] needs and ends that they serve, all belong to what technology is" (Heidegger 1977, p. 288). This means that technology does not consist merely of tools or products, but the whole social context in which the technology finds its use is part of what is technology also. Hence when one says that pre-modern technology is "enframing" in the sense of revealing the underlying truth, the meaning is that the technology, itself comprising the social and cultural context in which it finds its uses, shows that the whole society is also enframing and revealing. However, the situation is very different for modern technology. The dividing line between the two kinds of technology is that the energy used in pre-modern technology comes from either human power or the power of some animals, or forces of nature that is readily available. But in the modern world, we rely almost exclusively on energy that we dig up from underground, which we have to process a good deal before it comes in the form that we can use. The technology of the modern era thus becomes one of the *machine*. Instead of relying on sources of power such as the wind and the stream, modern technologists look at sources of energy from fossil fuels lying underground. Thus instead of minimally interfering with the course of nature when one puts up a windmill or the waterwheel to draw up energy, modern technologists have to tear up the surface of the earth and invent whole new ways of harnessing the energy contained in those fuels for the maximal use. Hence for Heidegger, the revealing nature of the modern technology, its "mode of revealing," is that it views nature as a "standing-reserve," (*Bestand*) something that lies ready to be used and exploited at any time. This is the crucial point in Heidegger's critique and analysis of technology. Modern technology is putting all of nature into its status as a "resource base," something that lies ready for use and consumption at any time. Moreover, the analysis goes wider; not only is the whole of nature taken as standing-reserve, but we humans are also being targeted and processed as yet another resource base too. We can see this quite clearly in the attempt by many websites to gather our data or our trails that we leave when we surf the online world. We are no more than a resource base, a source of data to be mined and categorized, by the giant software used by business corporations. In this sense we too are ourselves standing-reserves in Heidegger's sense. One is reminded here of the movie *The Matrix* where Morpheus is telling Neo about the true nature of the AI world and the resistance of humans. The AI is turning all human beings into their sources of energy; to the AI humans are nothing more or less than a piece of battery.

So what are the implications this has for the online self? Don Ihde remarks that Heidegger's critique of technology is intended for big machinery such as the hydroelectric dam, the ore extractor, or the nuclear power plant (Ihde 2009, p. 39). These are imposing structures, and they fit very well about their "enframing" power that makes everything a resource base. However, the online self is driven by information and communication technologies, whose hardware component is geared toward ever more minute miniaturization. We know, of course, that computers today are much more powerful than the most powerful computers one or two decades ago, but the formers consume much less power and take up much less space. This trend of miniaturization just about started to take shape when Heidegger wrote the *Question concerning Technology*. Ihde says that these technologies are even seen to be liberating and empowering, which stands in contrast to Heidegger's analysis of technology as constraining and dehumanizing. Instead the gadgets of information and communication technologies nowadays are often seen to open up possibilities and opportunities which were not conceivable before; some even say that they are empowering (Ihde 2009, p. 40).

However, the online self is neither hardware or software. It is more likely a product of software, at least its presence is made possible by some codes of instruction embedded in the memory section of the hardware. But it is not the software itself, in the same way that a poem written with a word processor program does not belong to the program. But this does not mean that Heidegger's vocabulary cannot be used to help illuminate the situation of the online self. The technology of the Internet is a revealing. It reveals the essence of modern technology, namely efficiency and optimization. The situation parallels that of Heidegger's critique of the typewriter as not a genuine writing, one that reveals the true essence of the humans who are expressing their thoughts through writing. Heidegger's writing on this topic deserves to be considered in full:

> Man himself acts [handelt] through the hand [Hand]; for the hand is, together with the word, the essential distinction of man. Only a being which, like man, "has" the word (μύθος, λόγος), can and must "have" "the hand." Through the hand occur both prayer and murder, greeting and thanks, oath and signal, and also the "work" of the hand, the "hand-work," and the tool. The handshake seals the covenant. The hand brings about the "work" of destruction. The hand exists as hand only where there is disclosure and concealment. No animal has a hand, and a hand never originates from a paw or a claw or talon. Even the hand of one in desperation (it least of all) is never a talon, with which a person clutches wildly. The hand sprang forth only out of the word and together with the word. Man does not "have" hands, but the hand holds the essence of man, because the word as the essential realm of the hand is the ground of the essence of man. The word as what is inscribed and what appears to the regard is the written word, i.e., script. And the word as script is handwriting.
>
> It is not accidental that modern man writes "with" the typewriter and "dictates" [diktiert] (the same word as "poetize" [Dichten]) "into" a machine. This "history" of the kinds of writing is one of the main reasons for the increasing destruction of the word. The latter no longer comes and goes by means of the writing hand, the properly acting hand, but by means of the mechanical forces it releases. The typewriter tears writing from the essential realm of the hand, i.e., the realm of the word. The word itself turns into something "typed." Where typewriting, on the contrary, is only a transcription and serves to preserve the writing, or turns into print something already written, there it has a proper, though limited, significance. In the time of the first dominance of the typewriter, a letter written on this machine still stood for a breach of good manners. Today, a hand-written letter is an antiquated and undesired thing; it disturbs speed reading. Mechanical writing deprives the hand of its rank in the realm of the written word and degrades the word to a means of communication. In addition, mechanical writing provides this "advantage," that it conceals the handwriting and thereby the character. The typewriter makes everyone look the same... (Heidegger 1992, pp. 80–81).

Basically put, the hand reveals the essence of the human being. Humans do not have hands, but hands are integral parts of the human from the beginning. The words, that is the human capability of rational thought and language, find their physical correlate not only in the mouth and the tongue, but also the hand. This gives a piece of handwriting a special status in that it is an extension of what it is to be a human being. On the contrary, the typewriter is something mechanical. It does not reveal what a human being essentially is. Thus, the typewriter "tears writing from the essential realm of the hand, i.e., the realm of the word. The word itself turns into something 'typed.'" That is, instead of the words appear as scripts which are essentially connected to the human being, an inalienable extension or even an integral part of the human being just like the hand is, the typewriter forces the words onto the sheet of paper in such a way that the human element is taken away. Thus Heidegger sees that the typewritten page "degrades the word to a means of communication." The typewriting makes all writing "look the same;" that is, all the individual quirks and characteristics that are there in the handwritten piece disappears when the same text is typed up with the machine. The essential human element is lost.

In the same vein, one can extrapolate this analysis onto the phenomenon of the online self. Word processing programs function much like the typewriter. There is the keyboard whose letter layout is taken directly from the old typewriter. The QWERTY layout was designed for the typewriter so as to slow down people's typing, instead of speeding it up, because speed typing would jam the hammers of the typewriter. One can see the letter one types on the screen, and importantly each piece of document written with a word processing program looks the same because the fonts and the machinery are always the same. The comments and posts on the social networking sites also require the same use of the keyboard as the typewriter. Even today's smart phones and tablets have keyboards most of which have the same layout as the typewriter. Hence, it is conceivable that, if Heidegger were alive today, he would have analyzed the social networking sites and the projection of the online self in the same way. If this is correct, then the online self is only a phenomenon where the essential human element is taken away. This could mean that the online self is at best an inauthentic persona, something fake that one puts up just as one is being constrained by the design of the software at various levels, from the operating system down to the browser and the sites such as Facebook themselves. Furthermore, this constraining is comparable to Heidegger's view of technology as enframing. In fact the enframing aspect is very clear when we consider that the online self appears on the computer monitor, which has its own physical frames. This is the only place where the online self can appear.

4.2 Jacques Ellul

Ellul is best known for his criticism of technology as an all encompassing force that threatens to envelop everything in its path, bending all components of society to be subservient to its own autonomous aims. What this means is that, once modern technology is let loose in a society, it will change all elements within the society so that everything works in accordance with its logic and its imperative. The logic of technology is the demand for efficiency, not only in the process of production, but in everything in life. For Ellul the product of a technology is a part of a large system that he calls *technique*. This is an all encompassing system that includes production process, distribution mechanism, consumption pattern, indeed everything in life and society in order that it becomes subject to the demand of efficiency and "optimization" as in management practice. This encompassing system is a necessary part of the technology, without which the technology does not work. Thus one should always look at a specific technology as a part of this system. A technology, such as the knowledge and skill required in producing a lithium battery, not only includes the knowledge of chemistry, but it is also enveloped in the system of use—what the battery is used for—as well as the kind of society that demands this kind of use, all of which is subject to the demands of the system of production that always stresses the need for best productivity.

Ellul's position is a clear example of technological determinism, the view that technology is autonomous and has its own dynamic logic that threatens to envelop every element in society to its will. The determinism forced by technology is, for him, almost total in that it leaves the people in modern society with little choice: Either they have to accept the path of technology or return to the pre-modern mode of living. Hence technology (in this sense we talk about 'technology' as 'modern technology' because pre-modern technology such as the waterwheel does not figure at all in Ellul's analysis) is an all or nothing affair. As returning to the pre-modern mode is not practicable in today's society Ellul's conclusion is that we have no choice but to accept the path that technology is forcing upon us.

Technological determinism has been largely discredited by most philosophers today. The view that technology comes with a large system that bends everything in society to its will is in fact not quite supported by empirical evidence. For example, a pure version of technological determinism predicts that when the Internet spreads to all corners of the world, it will largely homogenize those corners so that they will essentially become the same all over. One of the earlier beliefs in the impact of the Internet is that it will liberate the individualistic forces that may lie suppressed in some corners of the world, and those forces will be released with the advent of the Internet. The end result, so the prediction goes, will be that all corners will become the same, not only in outward appearances but more importantly in the mindsets of the people everywhere when they come to share the same set of basic beliefs, such as those concerning the value of individual choice and almost total destruction of traditional mores that have bound a people together in a distinct culture. Now everybody is free to do whatever they wish, and importantly they come to believe that *that* is a good thing. However, numerous research studies have shown that this prediction is not actually happening. Instead of homogenizing all cultures resulting in one big, global culture, what happens is that there is a hybridization of cultures so that the old and the new are mixed up in a contest. In some places such as China the traditional norm appears even to be winning as people seem to be content with the kind of Internet that is allowed by the government. That there is another kind of Internet consisting of China-specific websites such as Weibo and others (as substitutes for popular social networking sites such as Facebook, which are seen to be too Western oriented), shows that the prediction that the Internet will homogenize everything is not tenable. On the other hand, the Chinese Internet also does have its own dynamism, and is having very interesting impacts on the Chinese society in various ways.

That the prediction of the pure form of technological determinism goes wrong shows that the theory itself is suspect, but this does not imply that technological determinism is wholly wrong. A value of this view is that it looks at specific technologies as parts of a larger social system, such as the networked computer as part of the Internet and the Internet itself as a cog in the larger system of economic and historical globalization. This analysis has been useful in helping us understand various phenomena concerning the role of technology in society. What one needs to do is to modify technological determinism so that it becomes weaker and applies the analysis to more specific cases of a technology in a specific social context. For example, we could adapt the system analysis here for our analysis of the online self, and we find that the online self is itself part of the larger social, economic or cultural system. In this sense the online self can be seen as a manifestation of the impact that the Internet is having on the world. Hence the online self becomes a persona, or a front, that reflects quite clearly the globalizing force of the Internet itself.

4.3 Herbert Marcuse

Marcuse shares with Ellul the basic concerns for the possibility that modern society, through the work of advanced technology, would dominate social and individual life to the extent that individual freedom would be very difficult, if not altogether impossible. Both look at modern technology with apprehension and both stress the view that technology is an almost uncontrollable force that is bent on leveling and homogenizing both the physical outlook of social and cultural contours of the world and, perhaps more importantly, the mental and psychological outlook of its people. Ellul looks at technology as an expression of *technique*, which for him is a totalizing system that encompasses all aspects of life. Marcuse, on the other hand, looks at the issue through the eyes of a leftist. His main concern lies with the likelihood that technology will obliterate any form of effective resistance in terms of the kind of consciousness that "sees through" its dominating force coming in the form of the attractions of modern capitalist economy, so much so that the consciousness of the people would become, in his words, "one dimensional," a flat and featureless type of consciousness that leaves no room for any thought of finding genuine meanings and resistance (Marcuse 1991). Thus Heidegger, Ellul and Marcuse share the same view in that they largely subscribe to the typical view of technological determinism — technology representing an almost uncontrollable force having its own essence in the form of its inner dynamism and logic. All three also largely share the view that technology could be analyzed as one entity, so to speak; that is, as a symptom of the modern capitalist society that can be looked at as a whole. This viewpoint put them in contrast, roughly speaking, with the newer generations of philosophers of technology (see, for example, Ihde 2010).

What is most interesting in Marcuse's analysis for us is his view that the one-dimensional man has lost his functioning and effective private sphere. The dominating power of modern industrial society has resulted in him thinking the same way as his peer throughout the society. That is, everybody becomes subservient to the logic of the capitalist system. In his introduction to Marcuse, Douglas Kellner puts it very well:

> In Marcuse's analysis, "one-dimensional man" has lost, or is losing, individuality, freedom, and the ability to dissent and to control one's own destiny. The private space, the dimension of negation and individuality, in which one may become and remain a self, is being whittled

away by a society which shapes aspirations, hopes, fears, and values, and even manipulates vital needs. (Marcuse 1991, p. xxvii)

The image is that of a self that has lost his or her individuality, his or her power vis-à-vis the multifaceted power of the industrial society. It is the self that has her own agenda totally made up and brought to her by the consumer society. And all this is made possible of course by technology. As for the online self, one sees Marcuse's analysis when one sees that the online self is also subject to this homogenizing force. On Facebook we are all being bombarded by constant advertisements, whose content are carefully selected to attract our attention through our participation in the social networking world. In this sense, the online and the offline selves are not separated from each other; both are equally subject to this domination. Furthermore, as the online self is much more susceptible to changes and construction through words and images as it does not have a live body to resist those manipulations, the online self could well reflect very clearly Marcuse's analysis of the self here. As the offline self appears to be one-dimensional according to Marcuse's analysis, its online counterpart would certainly appear to be even more so, considering that it does not even have a body to anchor itself in the world.

Marcuse's emphasis on the depth of the mind and the self in Chapter Eight of his book (Marcuse 1991, pp. 207–228) is ultimately something that offers a redemption from the totalizing force of the modern industrial society. As much as analytic philosophy, in Marcuse's view, attempts to reduce the mind and the self to only linguistic constructions, these concepts refuse to be so reduced and there are always recalcitrant characteristics in these concepts that cannot be analyzed away. It is this resisting feature of the mind and the self and freedom that does not lend itself too easily to the reductive analysis of what Marcuse calls "analytic philosophy" that aims at leveling them to something that can be neatly manipulated through language and logic.

The foregoing makes it possible at least in theoretical terms the liberation of individuals from the grip of the industrial society. Marcuse has the following to say in his concluding chapter:

> Self-determination will be real to the extent to which the masses have been dissolved into individuals liberated from all propaganda, indoctrination, and manipulation, capable of knowing and comprehending the facts and of evaluating the alternatives. In other words, society would be rational and free to the extent to which it is organized, sustained, and reproduced by an essentially new historical Subject (Marcuse 1991, p. 256).

It is the "new historical Subject" that will be able to shake off the shackle that is put on by the modern industrial society. But the problem is how to find it and how exactly to characterize it. Marcuse believes that the new subject here would have to be multi-dimensional, in the sense that she is not reduced to the one-dimensionality that follows the path laid out by the system. Perhaps the new subject will emphasize her subjectivity and individuality. The key lies in basic mental orientation and consciousness. If the subject refuses to be led away by the modern technological society, then salvation is at hand. If the subject can be free from all propaganda and all forms of either crude or subtle brainwashing, then the subject becomes a new type, that is the new historical Subject. Marcuse thus sees his project as a contribution to the call to change the world. Rather than settling with simply describing and theorizing about the world, the new historical Subject would be much more active and more engaged in changing the material condition of the world.

An interesting question, then, is what form the online self is taking. Is it an old subject (supposing for the sake of argument that it can be regarded as a subject) which is one dimensional, or can it be the new historical subject? It seems likely that Marcuse would look at the online self as entirely one dimensional. After all, what could be more dimensional than something that appears on a computer monitor (which itself is flat, so two dimensional, but the point is the same) consisting of nothing but a bunch of information in visual, textual or audio forms? The online self can also be looked at as an expression of an advanced industrial, technology-permeated society where it is almost possible for the human self to exist in its entirety in cyberspace. However, the issue might not be as simple as it looks. It might be possible for the online self to acquire some amount of depth when the information that constitutes it is deep enough to include some form of critical analysis and resistance to the hegemony that comes with modern technology. For example, as social sites such as Facebook are increasingly being used to mobilize resistance force that aims at exposing the hidden hegemonic nature of certain political or economic agencies, the online selves function as the nodes on the site could then acquire some critical depth that at least might approach the threshold implied by Marcuse for the new historical Subject. In Marcuse's time some form of technology was used to aid the struggle for social justice and political freedom, such as the mimeograph machine and so on; thus it is conceivable that today's social networking sites can be used for these purposes too, and they are naturally much more powerful. As the online self can function as a persona for the person behind who puts the mask on in his or her engagement with their peer in order to achieve some social aim, the online self at least has certain capabilities of becoming deeper and more multidimensional than the perhaps overly simple analysis of it as necessarily one dimensional allows.

4.4   Albert Borgmann

In "So Who Am I Really?: Personal Identity in the Age of the Internet" (Borgmann 2013), Albert Borgmann presents a criticism of the Internet and especially the online self in an interesting way. For Borgmann the Internet is "a glamorous fog that has globally dissolved the contours of space, time, and people and at the same time condenses locally into brilliant if flat images of a place, a time, or a person." (Borgmann 2013, p. 18) On the one hand, the Internet literally exists in the servers and the cables and in the terabytes of information that is being transferred through various media every second all around the world; however, Borgmann sees that, on the other hand, the Internet transforms the very space and time that we live in. We are all familiar with the effect on time of the Internet (Hongladarom 2002;

Hongladarom and Kelly 2004). Scholars are familiar with the situation of having to wake up at odd times to participate in online discussion through the Internet with their colleagues half a world away. Many are constantly posting status updates on their social networking profile pages which show up on their friends' walls in the same way, no matter where the posts actually originate, either from an adjacent room or thousands of kilometers away. Time and space are being warped by the Internet. The impact of this warping, according to Borgmann, is that humans are losing their locality, or their authentic relationship with the immediate here and now. Borgmann's favorite example is the dinner table; instead of the family sitting together and engaging in face-to-face conversation with one another, now family members live in their own worlds with their individual smart phones. Or perhaps the family does not sit together at the same time at the dinner table any more. Even if the family is still together physically, it seems that each member is dispersing already, each one going separate ways according to whatever status updates each one is following.

The situation has an effect on the online self in various ways. Firstly, the online self has become a show piece, something that the user puts up in order to present their best aspect to the world. In this case the online self functions like a mask that people put on to show the world who they want the world to perceive. In this case the question "Who am I?" boils down to the online persona, which may or may not represent the real person sitting outside of the cyberspace. This separation between the online and offline self is a very important topic for this book, and we shall have a lot to say about this topic in this and other chapters. Borgmann's preference of the full dinner table is relevant here. Sitting at the family dinner table in the traditional way, with father, mother and children all present, is perhaps similar or comparable to face-to-face meetings of real people outside of the cyberworld. Here people can not only see and hear one another, but they can smell, feel, touch and hug one another too. Unless the information technology is so developed as to transmit smells or tactile feelings, this perhaps cannot be replicated in the online world. The dinner table becomes, in Borgmann's words, a "focal thing" (Borgmann 1984) which can deliver us out of the grip of the technological system. In the case of the online self, this would mean that one turns back from the fixation on presenting the online personae and return to the face-to-face basis of communication and closeness. The online self is but a front-end of a vast system of the Internet much of which is hidden from us ordinary users. Even what is going on inside the computer with which we are familiar with is a total mystery to most of us. It is this overarching system, Borgmann tells us, that is the culprit. The system engulfs us and makes us unable to see the reality as it really is, that is, as something which we can touch and get really close to. It is here that the online self represents for Borgmann perhaps the epitome of this technological system. What is happening here is could be compared to Heidegger's view of the electronic resource functioning as "standing-reserve" (*Bestand*) where the electronic component that makes up the online self works as a resource that lies ever ready for the user. The overarching system of hardware and software is thus comparable to the technological system working as something that "enframes" us, making us unable to see other possibilities that lie outside the system and thus constraining our thoughts and consciousness. The system of the Internet thus enframes us so that we come to believe that the online self is continuous with our own, flesh and blood self. It is something that is always pliable, something that we can mold at will in order to present what we want to be perceived as to the outside world.

This condition leads to the second of Borgmann's critique of the online self. Here identity is reduced to the information that is available in the "About Me" section in a website, or in the profile page on Facebook or Twitter. Personal identity is then little more than some glitches on the monitor showing attractive photos of the subject, one which the subject carefully chooses to present her best look, together with some information about the subject herself. The information here differs from the typical information one finds in a paper resume in that in the latter the purpose is to present information at a more concrete level. This is similar to Borgmann's theory of information and reality where he divides the relation between the two into three categories, namely reality as information (as in smoke signals, or in natural information such as smoke signals fire), information for reality (as in speech or writing systems, whose meaning signify outside reality), and information as reality (as in digitized images where the image, constituted by information, becomes reality in its own right) (Borgmann 1999). The paper resume thus works at the middle level of information for reality. However, the online profile on Facebook appears to function according to Borgmann at the most abstract level, that of information *as* reality. For Borgmann this is something that should be resisted because that would mean that we are further and further estranged from bedrock reality. The online self would mean for Borgmann that we are also estranged from the kind of reality we are familiar with, in this case our own self. In addition to the "offline" self that we are familiar with, information and communication technologies present us with the possibility of creating, on demand, information as reality, something existing only in the cyberworld and capable of taking on as reality on its own. In this sense our situation is markedly unlike that of the medieval monk Suger whom Borgmann talks about in his paper. Suger's identity is fixed through his birth, his town, his family, and so on, but our situation is much more fluid. Not only can we change our identities at will (or so it seems), but Borgmann seems to believe that we can also create our own identities in any way we like. An example might be given that it is not uncommon on the Internet to find a man posing as a woman in the online world, and vice versa. This situation is totally unconceivable in the medieval world.

4.5 Don Ihde

In *Bodies in Technology* (2002), Ihde presents some contrasting features between the real life body and the virtual body, a discussion that bears some similarities with our present discussion on the online self (Chap. 1). Ihde starts by talking about two types of experiences, one embodied and the other disembodied. He asks his students to tell him about their imagined experiences, something that they have not had before, such as jumping out of an airplane in a parachute. Roughly half of the students tell their experience from the first-person perspective; that is, they tell of how they feel the wind rush by their faces and bodies, how the ground appears larger, how the opening chute pulls up their bodies, and so on. The other half, on the other hand, tells of their experiences from the third-person. They imagine themselves falling down to earth with the chute, but they imagine seeing themselves from a certain distance, as if they themselves, in the first-person, are floating nearby and watch themselves falling through the air. Ihde calls the first kind of experience an embodied one and the other disembodied. Ihde is discussing the worry that many are having about the onset of virtual reality technology. Is real life being replaced by virtual reality? Is the line or the boundary between the two disappearing so that we cannot tell which is which anymore? Is the human body being replaced by the machine? Ihde's another example is that of someone who has an out-of-body experience. She is experiencing seeing herself lying on a table, whereas "she herself" is floating around the ceiling looking down on herself. Ihde calls the body on the table the "image body" and the one who floats the "here-body" or the "me-now." (Ihde 2002, p. 5). The fact that the floating body can see the image body on the table as being oneself is for Ihde an indication of virtuality. The image body in this sense thus becomes the virtual self, something other than the body or the self that is doing the seeing and thinking.

This situation can then be compared to that of the online self where the latter becomes the image-body, one that is always perceived by the here-body. This is perhaps akin to the situation where the user, who has her offline self, watches and edits information constituting her online self or persona. The difference is, of course, that the online self can "talk" or provide information by other means, as one can see when somebody writes a comment on Facebook. Each comment box contains a link to the profile page of the commentator, in addition to his or her profile picture. So it is as if the online self can actually talk because usually the language in these comment boxes is highly conversational. We know, of course that somebody in real life is typing up these comments, but the appearance onscreen seems to indicate otherwise. Similarly, we can also imagine a situation where the image-body walks and talks, but since the body here is only an object of the perception of the here-body then the former is not in a position to replace the latter totally. We can watch our double, as image bodies, do a lot of things, but we cannot *be* them because we are located here looking at them from the outside. Thus Ihde does not believe that virtual reality can totally replace real life.

Even though the here-body is augmented by technology so that it becomes immersed in virtual reality experience, there are still gaps wide enough in it so that the subject feels disoriented (Ihde 2002, p. 11). So for Ihde the experience in VR programs is only theatrical; it functions as real life as well as a theater can (Ihde 2002, p. 11). For Ihde the relation of the self to technology is always ambivalent. He sums the view up here very well in the following passage:

> The direction of desire opened by embodied technologies also has its positive and negative thrusts. Instrumentation in the knowledge activities, notably science, is a gradual extension of perception into new realms. The desire is to see, but seeing is seeing through

instrumentation. Negatively, the desire for pure transparency is the wish to escape the limitations of material technology. It is a Platonism, returned in a new form, the desire to escape the newly extended body of technological engagement. In the wish there remains a contradiction: the user both wants and does not want the technology. The user wants what the technology gives but does not want the limits, the transformations that a technologically extended body implies. There is a fundamental ambivalence toward the very human creation of our own earthly tools (quoted in Ihde 2002, pp. 13–14).

Thus, for the online self, the ambivalence is perhaps where there is a tension between, on the one hand, the feeling of elation and being empowered one feels when one projects oneself onto the online world, when one is capable of connecting to thousands of people all around the world at the same time. On the other hand, one feels the limitations imposed by the technology when the online self is nothing more than a collection of pixels and digital data. The digital self one has on Facebook is not exactly the same as the self one feels that one has in the virtual reality programs. The former is a projection in the outside world—by logging on to my Facebook account, I am not thereby renouncing the world outside and enter into a virtual world, being immersed on all sides by the goggles and other devices. On the contrary, I am awake and engaging with my friends in real time, many of whom are commenting on very concrete and actual issues that are happening in the real world outside. Here Facebook and other social networking sites are clearly continuous with the outside world. This does not have to happen in VR programs.

Ihde's notion of embodiment relations is also a useful tool with which we could understand the online self. For Ihde, "*[e]mbodiment* is, in practice, the way in which we engage our environment or 'world,' and while we may not often explicitly attend to it, many of these actions *incorporate the use of artifacts or technologies*" (Ihde 2009, p. 42). We have seen that for Heidegger the hammer becomes an extension of the self of the user when it merges seamlessly into the user's work, just like the hand is. In this case the hammer is not specifically being paid attention as a hammer but the focus of the user is elsewhere, such as, for example, with hitting a nail in the process of making a table. The hammer then exemplifies the embodiment relations; it is as if the hammer has become part of the self or the body of the user, again just like his or her hands are. In this sense, then, the online self thus becomes part of the embodiment relations of the user when it ceases to function as an object of the focus of experience by the user, but functions transparently when the user works through the online self when she is engaged with a task. The difference between the online self and the hammer is that the latter is a tool, and when it is ready-to-hand it represents only a part of the self, one that does the hammering work only. The online self, on the contrary, could function in a more comprehensive way, such as representing the whole of the self of the user, including her personality or a constructed persona entirely.

One main contribution of Ihde is that he suggests a postphenomenological turn where the attention of the philosopher is not on technology as a monolithic entity subject to phenomenological analysis, but as a general term denoting a variety of technological products or processes which do not need to lend themselves to become merely instances of one overarching concept. Another way of characterizing the postphenomenology is that it is a non-foundational and non-essentializing phenomenology (Ihde 1993, p. 7). Hence, instead of going Heidegger's way and focusing on Technology (with a capital T, implying, of course, that it is one big monolithic entity), the postphenomenological turn focuses on the specificities of technologies, such as the automobile, construction of bridges, development of optic technologies or hearing aid technologies, and so on. One of Ihde's main criticisms of Heidegger is that he fails to anticipate the development of technologies after his death. He could not have foreseen the advent of the Internet or the smart phone, for example (Ihde 2010). These technologies contain so many important specificities that the broad stroke that Heidegger offers is no longer adequate. There is no assumption that something foreboding or threatening ties these technological processes or products together. In this manner Ihde's postphenomenological turn shares something with sociology of technology; however, the critical viewpoint is still there. In short, the postphenomenological turn consists in paying attention to findings from related disciplines such as history, sociology and other related fields in order to focus on the specificities surrounding a technology and to shed light into the phenomenon of technology in society. Rather than thinking about technology in the abstract, employing ancient Greek terminology and searching for an essence of technology, Ihde is doing what many philosophers (mostly those of the younger generation) are already doing. That is to say, they look over their shoulders to find out what their colleagues in related fields are doing and adopt those findings into their philosophical analyses. Ihde contends that Heidegger was blinded by his insistence that the essence of technology is obscured by the particularities of separate applications of technologies. But in fact it is those particularities that are much more interesting because they represent the actual cases of how we human beings interact with the world in a way that reflects our desires, goals and attitudes which are relative to changing circumstances.

4.6 Hubert Dreyfus

Dreyfus is another of philosophers of technology in the Heideggerian tradition who is rather critical of technology as a whole. His earlier work, *What Computers Can't Do* (1978), is a strong criticism of the AI program in computer science. His main contribution to the critique of the role of Internet in the human worlds appears in an article, "Anonymity and Commitment of the Internet" (1999), where he discusses the role that online learning plays in education. Basically his argument is that the Internet makes it much easier for students to "shop around" so to speak, sampling this or that content with the result that they are not really committed to any program or any content in particular. Dreyfus compares the situation of the Internet with Kierkegaard's criticism of the modern society where its members content themselves with reading the newspaper, consuming news made up by the media and thereby becoming only a mere cog in the machine of society without asserting their own individuality. The picture that Kierkegaard paints here is that of a person who believes that he is a "good" member of society, one who pays his taxes, does his duty to the community, works hard to earn his living and reads his news to keep abreast of what is selected for the mass to know and to be on the same wavelength as their social peer. For Kierkegaard, however, this kind of life is not at the most developed level that a human being could achieve. It is merely at an *ethical* level, where the person is expected to conform to social rules. The rules themselves might be rationally justified—that is precisely the point of its being ethical, but for Kierkegaard a life at this level leaves something out and thus is not fully satisfying. Following all the rules results in the person becoming indistinguishable from his or her fellows. Viewed from the perspective of Kantian ethics, for example, an individual who deliberates on what to do and what not to do has to follow a set of universal maxims, or rules, and following these rules is only moral when the rules are applicable to everyone without exception. Thus the uniqueness and distinctiveness of the individual is not taken into consideration in the Kantian system. The individual's existential condition, her unique characteristic that separates her from all other things in the universe and more importantly her unique properties that qualify her as a unique human being, is lost in Kant's system. For Kierkegaard this is unsatisfying because he believes that there is another dimension to human life, one that pays very specific attention on the very uniqueness of an individual in her capacity as this individual and none other. Kierkegaard calls this level the *religious* one. The religious way of living transcends the ethical way in that it aims at something beyond the way of living according to reason, that is, according to the rules laid out by reason and under the presumption that everybody is the same. The religious aspect of life is predicated on the notion of faith, and for Kierkegaard faith is central to a full human life because it means one is always prepared to "jump" across the chasm between reason and unreason and is ready to accept the latter. For Kierkegaard this readiness to jump, a leap of faith, is a condition for very deep commitment that one can have toward whatever is object of one's faith, and this is something that is lost in modern society.

Dreyfus bases his criticism of the Internet and online learning on this apparent lack of deep commitment that is only possible when one assumes the religious way of life here. The Internet makes it much easier for the individual to float around, sampling this or that piece of information without becoming really passionate about anything in particular. Dreyfus compares the average user of the Internet with Kierkegaard's newspaper reading public, who can form opinions on a variety of issues but themselves are not directly engaged with any issue in particular. Dreyfus sees that Kierkegaard's Public as detaching themselves from taking responsibility in any exigencies that might be reported in the newspaper. Instead of the members of the old polis who take full part in the administration of their community, the newspaper reading public assume the role of spectators who watch the affairs of their community without actually becoming players. Dreyfus sees that the Internet only makes this situation more serious because of its ubiquity and its very effectiveness in disseminating information. The self in the age of the Internet then becomes at most an ethical self in the sense described above, or an *aesthetic* one where the main concern is mostly consumption of entertainment and personal enjoyment. Dreyfus sees that it is very difficult, if not impossible, for the average member of the public to rise above either of these levels and achieve the religious one where there is full and deep commitment.

One can then see how Dreyfus would have regarded the condition of the online self. At most the online self would be only at the ethical level; it would be very difficult for the online self to achieve the religious level of existence because it exists entirely, or so it seems, in the online world, a world which for Dreyfus is comparable to, or is an extreme version of, the generic newspaper that Kierkegaard criticizes. The problem thus is whether the online self can achieve the level of commitment that would have satisfied Kierkegaard or Dreyfus. One way of looking at this question is to ask whether it is actually possible for someone to be true to her existential condition and to have a strong faith in the sense described by Kierkegaard while at the same time being fully engaged with the world of information technology and the Internet. If this is possible then the online self does not have to be prevented from achieving the kind of commitment that Kierkegaard talks about. One of the interesting phenomena in the cyberworld is the use of the technology to gather support for a cause and to mobilize people for political purposes. Websites such as Change.org offer online petition tools where an average user can start a petition and gather support from members of the public on any topic. It is imaginable that the person who starts a cause on, say, saving tigers from extinction, can be very passionate about her cause. She might work directly with tigers in the wild and know the natural and political conditions surrounding them very intimately, and she might well believe that without awareness and strong support from the public, the tigers might become extinct in the near future. So she starts an online petition called *Save the Tiger* and aims to gather enough support to move the relevant political authorities into action. She is very committed to the cause and is very passionate about it and through the online petition tool this commitment and passion can spread to others whom she does not know personally at all. So does this qualify as Kierkegaard's level of commitment? We can further imagine that her faith that tigers must continue to have their place in the wild could border on being religious. We could also further imagine that she produces a leap of faith where her zeal for saving the tigers transcends the ability of average reason to comprehend, and it is this zeal that spreads to other members of the public who signs up for the petition when they see it on the Internet. On the surface at least all the conditions required by Kierkegaard appear to be satisfied. The originator of the petition is very committed; she is not anonymous and neither are those who sign up because everyone uses their real names; her zeal for saving the tigers is at least equal to those with religious zeal. If this is the case, then we may have an example where there is a level of commitment and zeal in the online world that would have satisfied Kierkegaard.

However, Dreyfus might counter that in this case the one who has the commitment is the originator of the petition herself, but the members of the public who sign up their names do not. What they do is that they read about the petition which comes to their attention usually through the social networking sites or emails and then they feel that they agree with the content and then they sign up their names. This does not have to show that they have the same level of commitment as the person who actually works with the tigers. However, if the petition succeeds in forming a group of people who decide to meet in real life in order to push forward their agenda, it is possible that this group should have at least some level of commitment to the cause in a way that extends beyond only reading the web content and typing up their names. Cases where groups are formed out of members who come to meet one another in real life through information provided and shared in the online world show that the online and the offline worlds are quite closely integrated and those members of the public who consume the news on the Internet do not have to confine themselves to watching the computer monitor and typing on the keyboard all the time. There can be some real commitment among the members of the online public too.

4.7 Andrew Feenberg

Feenberg's main contribution to philosophy of technology is that he introduces a critical study of technology that pays attention to the role social forces play in designing and construction of technology (Feenberg 1999; Feenberg and Friesen 2012). Instead of technological determinism that we find in Heidegger and his followers such as Borgmann or Dreyfus, Feenberg looks at technology as a dependent variable, one that can vary according to social or economic contexts. Here Feenberg relies on findings by social scientists such as Bijker (who talks about the design of the bicycle) who investigates the role these socio-economic and historical forces play in the direction in which specific technological products are heading (Bijker 1995). Instead of looking at technology as a finished product, one where the onlooker feels helpless as to how and why it appears on the scene, Feenberg and many other philosophers of technology tend to look at technology at a more inclusive level. They tend to look at the stages in which a specific technology is conceptualized, designed, tested and brought about to the world. This is a standard way the process of technology design is studied; its importance is that by looking at how a specific piece of technology is conceptualized and designed, the technology ceases to become a threatening force but something that one is involved in and understands from the beginning. The philosopher, one who reflects on the role and value of technology in society and culture, thus feels that she is not always at the receiving end, but can have a say in how a technology is designed even before it comes out as a finished product to the market (Ihde 2004). This sociological investigation thus supports a thesis that technology both shapes up and is shaped by other elements within society, and that there can be many possible directions that the technology can take depending on what kind of choices the designers take which in turn can be influenced by the values and goals that society has at that particular time as to what the technology in question will be for. This thesis, known as constructivism, thus undermines the broad position of technological determinism which takes it for granted that the path led by technology appears mostly to be one way only.

Feenberg is also well known for his view that the direction of technology design, once understood to depend on the values and goals that people take, should then be governed through democratic means. Adapted to the phenomenon of the online self, that would mean that the process by which the online self comes to the fore should be a democratic one. Perhaps the way Facebook is run should be democratic in the sense that the users should have a say in how the website appears and how its various features should be employed. In face Facebook already has instituted a kind of user input when they try to introduce some new features, such as a change in how the privacy settings of the users are affected. Moreover, the democratic control envisioned by Feenberg can also be seen in another way such as when the online selves—the people who are connected to one another through the social media—get together to push forward their political agenda, either on the social media or outside in the offline world. The phenomenon thus reflects Bakardjieva's view of "subactivism" where the ordinary users are too preoccupied with their day to day living but still they want to participate in politics so they use the online groups to achieve their objectives (Bakardjieva 2009). If the subject matter of the getting together concerns how technology should be used in the public arena, such as whether a hydroelectric dam should be constructed on this or that river, then this illustrates Feenberg's view of the democratic control of technology. In either case what is salient is that the boundary between the online and the offline worlds are becoming more and more artificial.

The issue here touches upon a very important problem in the discussion of the online self—the problem of agency. We shall deal with this topic directly later in the chapter. In any case, Feenberg's analysis helps us see quite clearly how the problem of agency becomes an important one in our attempt to understand the online self. As constructivism shows, technology does not come down to us from on high. We ordinary non-technical people can do more than merely receiving any form of technology that is handed down to us, but we have the ability to join force and direct the way a technology is designed and used too. The "we" here quite naturally includes online selves too, especially when we consider that much of the deliberation on public policy issues nowadays take place in online arena. The problem, in a nutshell, is: How can online selves, who are apparently constituted through information in form of bits and pixels on the screen, be in fact active agent? If the online self assumes the status of more than just inert personae for the offline users, then it has to have some ability to function as an active agent. But how is that possible at all? Feenberg's contribution in our investigation here is that his analysis of constructivism and his critique of technological essentialism bring this problem to the fore.

Viewed through the debate between essentialism (or determinism) and constructivism (or critical theory of technology), the online self is a representation of either an effect of the domination of efficiency or the consumerist attitude on the one hand, or an opportunity that one could take to engender social change on the other. The crux of the matter lies in whether the online self has a choice to make the decisions that she believes she is making (and, of course, whether it makes any sense at all to talk about an *online* self making a decision). Here the analysis boils down to the question whether one has a real choice in the environment of the advanced capitalist society that is saturated by technology. This analysis, then, becomes embedded in the debate between essentialism and constructivism in general. This is also in line with the idea that the online self and the offline one is continuous to each other and an analysis of the one usually is applicable to the other.

4.8 Agency

Feenberg's analysis helps us see clearly the relations between the online self and its various contexts, especially socio-economic and historical ones. Not only is an analysis of the online self as to what it is in itself quite limited, it also is inadequate considering that a fuller understanding of the phenomenon cannot actually be obtained if these various contexts are not taken into consideration. Thus, for example, the online self should not only be investigated as to whether it is constituted by information or whether its narrative can account for its identity, but we would miss quite a significant amount of understanding of the role of the online self in society if it were to be considered alone in itself. This is where the critical perspective comes in. Here the online self is considered both as actor and actee of social concerns. Political and economic considerations do play a role in how the online self in general appears the way it does, but the online selves by themselves can affect real social change in the offline world too, something that Bakardjieva has done an extensive research about in her work on subactivism as we have seen. This leads us to what we have earlier discussed but had to postpone until now, namely the problem of agency. Basically put, the problem is how we are to understand the phenomenon where the online self apparently becomes an agent who leaves indelible marks in the real world. This problem is closely related to the one on responsibility. An online agent does something that produces real changes. As agent she must be responsible, but how can we hold an online agent to be responsible or liable? How tenuous is the relation between a user and his online persona? The problem has become very important today due to the increasing threat of online theft. Identity has become highly valuable; it has become a closely guarded commodity but has often been stolen using more and more sophisticated methods. There are numerous works on identity theft and various aspects of online behaviors. My concern here, however, is a philosophical one: What accounts for the identity between an actual, real world user and her online self? If an online self can actually be considered a self in his or her own right, then how could the self be held legally liable if she did something against the law? Since the line between the online and offline worlds is artificial anyway, a rough and ready answer could be that the real world user who manipulates the persona should be the one who is responsible. But, metaphysically speaking, what exactly is such a connection? How are we to understand it?

Recall that our thesis in the previous chapter is that the self is constituted by information and that it can and does extend outside of the skin. This basically solves the problem of agency at a stroke. In this sense the online self and the offline user are one and the same, as they constitute each other; hence there is no problem in principle of linking the two when offline accountability for online action is required. As things stand at present, online selves have no brains, thus they are not conscious so they cannot by themselves be agents. But if technology progresses far ahead so that computers and humans are fully merged into each other, we might imagine a situation where the biological brain is replaced by the neural network of the computer. In this situation the "cyberworld" does not exist only on the monitor, but real life does become cyberworld and vice versa. In this scenario, projection of the data processed by the computer does not limit itself to the computer screen, but what we normally experience as three-dimensional reality could itself be such a projection, one where there are people whom one can talk too. In this case there is no reason to think that online selves cannot become conscious by themselves. Hence the view that online selves can eventually become conscious on their own is not as farfetched as it might look at first. In this case, then, the problem of how to get someone to be accountable for an action would be to find the person who conceives of the action and consciously carries it out. Such a person does not have to be a biological being, but could be made of other material, such as silicon. If the silicon-based being is conscious and is a rational agent, then he or she has to be accountable and responsible for his or her action. If this can be the case, then, the problem of how online selves can become conscious, autonomous agents will not sound impossible after all.

We can then imagine further that these machine-human hybrids can get together to engage in a political action. Perhaps they want to protest for equal rights with the traditional carbon-based human beings. They can also engage in Bakardjieva's subactivism when their main preoccupation is elsewhere and they join up the protest group through their "online" channels which do not occupy most of their time. In fact we do not have to imagine a scenario where the online and the offline are fully integrated like this to see the point of the problem of agency, even at the moment when the online exists almost exclusively on the computer screen we can also see that the problem of how to account for the agency of the online self boils down to the agency of the user in the real or offline world.

In the literature on information ethics there are works discussing the problem of agency in artifacts (Floridi 2010). For example, if a driverless car happens to bump into somebody, whose fault is it? Is being injured by the car similar to being injured by a working machine where one has to know beforehand that it is dangerous to wander close to the machine in the first place. Or is it similar to a case of being bullied in an online situation by a cyberbully, where, of course, the bullier is fully responsible? Scholars in science and technology studies investigate the role of the system in which an artifact derives its function and meaning and try to see how the system has any role to play in the accountability or responsibility of artifacts (Wiegel 2010; Allen 2010; Floridi 2010). According to Floridi, even though it is too early to claim that machines or artifacts can be responsible for their action since that requires conscious action, they can be held accountable (Floridi 2010, p. 88). A rescue dog cannot be responsible for her action, but since she usually receives praises when she does a good job, then Floridi sees that she deserves praises from the people who are grateful to her work. The praises are meaningful because we appreciate the work of the dog when she rescues people who are trapped under collapsed buildings. We know that the dog is not concerned with compassion or empathy or the understanding that there are people who need to be saved—for her she is just doing what she has been trained to do, but we praise and feel thankful to her anyway. In this way she is accountable to her job, according to Floridi. Similarly, the driverless car that bumps into someone, presumably, would itself deserve some blame for its action, and the car that successfully delivers its passengers to their desired destination would deserve some praises in the same way the rescue dog does.

In fact there appear to be a number of problems with Floridi's distinction between being responsible and accountable here. Floridi uses the example of a rescue dog, who usually gets rewarded and a lot of praises for her work. However, it is a well-known dog psychology that constant rewards and praises are an important factor in training the dog to do whatever we want them to do. Dogs are by nature creatures that try to please humans all the time, so giving them rewards and praises reinforce their desired behaviors, such as looking for people buried in rubbles. We praise the dog even though we are fully aware that the dog does not understand that it is ethical to save a human being who is trapped behind collapsed buildings. The praise is not given to be dog because the dog is a fully conscious and autonomous moral agent, but it is given because we would like her to continue doing what she is doing. The dog may have some level of consciousness and, some might argue, some understanding of what is going on, such as human beings need to be pleased and by rescuing them from underneath piles of rubbles please them immensely. Some dogs even get medals for their heroic effort. If they can be regarded as having only a certain level of consciousness and, perhaps, belief, then to reward them would not be too far off. On the other hand, if a fully autonomous, driverless car successfully delivers its passengers on time, safely, and efficiently, we naturally would not praise or reward the car in the way we reward the dog, simply because we do not think that the car knows anything about what it is doing. Instead we do praise and reward the inventor, designer and manufacturer of the car instead. Yet for Floridi he seems to take the dog example to be representative also of cases where 'responsibility' is distinct from 'accountability.' The car and the dog may indeed by in the same category, that is, the category of non-human moral agent—but since the dog is closer to us biologically than the car, Floridi's use of the dog as an example thus appears to be designed to lead us to believe that such a distinction also makes sense for non-biological beings such as the driverless car too. But since we do not typically reward cars, but their designers, the dog example perhaps does not carry over into these non-biological artifacts. Consequently, we have to find another example or another source to justify Floridi's distinction here.

All these are directly related to the online self. If in an online setting somebody does something unethical, such as when he or she poses as a respectable character, solicits donations from other, unsuspecting, people and disappears, would he or she be responsible for it? More specifically, if an online self does the same thing, is the online self itself (or himself or herself) to be responsible? What if the character here is in fact a robot designed to pose as a respectable character soliciting donations? Who is responsible? The designer of the software robot? Or the robot program itself? According to Floridi, the robot should be held accountable, but it is not responsible. Taking the dog analogy, perhaps the robot should be blamed, perhaps rebuked strongly so as to discourage from doing the same thing again, just like what people do when they train a dog not to repeat a negative behavior. But this does not seem to be effective for a robot, unless the robot is wired with a program that helps it learn through reward and punishment. That, however, is still far in the future. Thus, if the online self is only a persona of an offline user, then there is no problem of agency and responsibility. The user is liable for his or her action, and the problem boils down to an empirical one of tracking who the real user is behind the online persona. And if the online self and the offline self are more closely integrated, as in my argument, then the case is even stronger. In case where the online self is the work of a robot program, then the issue is a little more complicated. Instead of blaming the robot for its negative behavior, we should hold the one who designed and released the robot on to the Internet. If someone with a malicious intent designs a robot software aiming at creating havoc on the Internet, then that person is held accountable and responsible for her action. This is quite straightforward. But if the intention of creating the robot program is not malicious, but the robot fails to function as designed, then the designer is still responsible, but for the failure of his design and not for the malicious intent. In the case of the driverless car that bumps into people, the designer or the manufacturer should be held responsible for either faulty design or faulty manufacturing. The issue hinges on whether the online self, as an extension of a larger self which may be a large computer program, is even minimally conscious or not. If it is not, then even to attribute what Floridi calls "accountability" would seem to be quite farfetched.

But what would we do if the behavior of the online self is more an "unintended consequence" rather than something that has been deliberately designed? What if the designer did not have a malicious intent, but since programming a software robot is a hugely complex task, there is always a chance that the robot and its online persona or self could perform tasks that which they are not actually designed for. In this case the situation is structurally similar to the driverless car that accidentally bumps into people. The designer of the car does not have any malicious intent, but the car somehow loses control and creates the accident. Many STS scholars attribute agency in cases like this to the structure or the system within which the technology operates. For example, the system could be such that it gives rise to a need for a driverless car. The system may be rather unforgiving, one that drives the designer as a cog is driven in a giant machine that is pervaded by the idea of efficiency. We then return to the problem of technological determinism again. STS scholars would point out that in such a situation even though no one in particular is responsible for the accident caused by the driverless car, it is the system itself that is ultimately responsible (Wiegel 2010). We could compare this to a social arrangement where women are treated as second-class citizens. The men who regard and treat women in this way may not have malicious intent—they do not intend to harm any particular woman in any way, for example, but the overall social system is such that both the man and the woman and everyone within it are victims of the arrangement that promotes this kind of classification. The only way out, of course, is to see through the oppressiveness of the system and try to change it both from within and without. A system that drives people to work too hard would make it easier for accidents to happen, so when accidents do happen instead of just accepting them to be accidents a culprit can be identified which is the system itself.

To be more specific, when someone is being deceived by a robot program posing as a human online persona, the program itself should not be held either responsible or accountable, even in Floridi's sense. This does not mean, of course, that no one is going to be held legally liable for the scam. Common sense tells us that it is the designer of the software, the one who planted the software on the Internet with the malicious intent of scamming unsuspecting user, who should be held liable. In this sense there has to be a way of linking the online persona with the real user, which is an empirical and technical matter. Nonetheless, there are examples of autonomously working software that creates huge impact such that if the software were human agents they would certainly be put in jail. The 2008 economic collapse that took place in the US was believed to be at least partly caused by autonomous software that automatically "decides" when to invest in certain stocks and how much. When something unexpected happens, these programs continue to pour millions of dollars into underperforming stocks, causing a worldwide financial disaster. According to Floridi, the investing software here is called "artificial agent" (Floridi 2013), which to him is capable of acting in such a way that it is accountable or responsible for its own action in one way or another. According to Floridi, the software agent here is part of an ecosystem that he calls "infraethics," namely system or environment that makes smooth communication and transaction between users possible. Infraethics thus deals with the right sort of infrastructure that enables action among users which can be either moral or immoral. For example, good roads can be used for transporting lifesaving medicines, or they can be used to carry life threatening weapons. The road is part of an infrastructure and the idea that good roads should be maintained is part of infraethics. Here one can see the clear role of information and communication technologies because they are the technologies that serve as ecosystem or infrastructure that make virtually all kinds of interactions possible (Floridi 2013). Floridi is optimistic in that he believes that the benefits of the ICT infrastructure outweigh the risks (Floridi 2013), the stock investing software program perhaps not withstanding.

Floridi ties the idea of infraethics up with his view on distributed morality. The latter is the morality of collective action, such as when millions of credit card users allow the credit card company to deduct a tiny percentage of their transactions in order to help support certain good causes, such as providing medicines to AIDS patients in developing countries. He sees that distributed morality includes the artificial agents such as online programs and so on; the key idea is that infraethics is needed to provide a guideline to a kind of action that leads to a well-functioning of distributed morality. In his system of ethics, he tries to downplay the role of individual users who could be held responsible for such and such action. If some objective good emerges in the overall ecosystem (what he also calls the infosphere), then that is ethically positive. His view here thus resembles that of Spinoza, who argues that ethical action is what leads to objective congruence or well-functioning in the entire reality (Spinoza 1985; Hongladarom 2008). What matters is not so much that one or more persons have the right intention, but that in the entire ecosystem some objective good is increased or not, and for Floridi the amount of information available in an infosphere represents such objective good (Hongladarom 2008).

In this environment, then, the online self is thus an aspect of distributed morality facilitated by an infraethics that he mentions. Presumably the physical and software infrastructure that enables the online self to be there is the matter of infraethics, and whether the online self does something good or bad (such as trying to scam others or inviting other to join some good causes) depends on this or that particular online self himself or herself. By downplaying the issue of responsibility, then, Floridi does not directly address our concern here. Even if we accept his view about infraethics and ontological ethics in the infosphere and infraethics and so on, we are then left with the question of how to account for the responsibility of the action of online users. For Spinoza, a good action is one that is in accordance with the rational structure of reality; that is an action is good to the extent that it promotes the rational structure or not. We can readily compare Spinoza's rational structure of Substance or reality with Floridi's insistence on the amount of information in the infosphere here. What happens in an individual's mind is not so important when the focus on ethical concern is on the outside, so to speak. In this sense, then, there is little that is surprising when Floridi argues that non-human and even non-conscious machines can become ethical; they are so because they can certainly contribute to the increase of the overall amount of information, which for him is ethically good in itself.

However, Floridi also talks about cases where there is not one single person who is responsible for unethical or illegal action on the Net, but the action is a work of a committee or groups of persons (Floridi 2013). In this case common sense also tells us that the whole group must be held liable, just as when a whole gang can be arrested and put on trial if evidence shows that the gang actually committed a certain crime. In the case of the online self, when a large number of crimes in a smaller scale are committed by a large number of online selves, each contributing their share to the collective crime, the situation should not be too different from the crimes committed by people in the offline world. If a number of online selves participate in sharing some illegal material, perhaps child pornography or instructions on how to make homemade bombs, then the harmful consequence will accrue to the society of those selves as a whole even though each contributes a small part to the task. In this case each of them is responsible. Thus we see that the online self is naturally responsible, unless that self is a persona of a robot software, in which case the designer and the one who puts the system into operation must be responsible. In the future case where the software itself becomes conscious, then certainly the software is responsible (the software may be housed in a bodily form such as a live robot machine). Or when the collective itself becomes conscious, much like when somebody says that the beehive itself can become conscious or a "superorganism" over and all above all individual bees in it, then the collective is certainly responsible. Even if the software itself is not conscious yet, it can be regarded as taking a share of the responsibility when it functions as a part of an overall system. Furthermore, if one subscribes to the view that the system here can work autonomously (such as the view of determinists such as Ellul), then the system can perhaps be regarded as being responsible even though of course in itself it is not a conscious being.

4.9   Continuity

The implication of the previous chapter on our understanding of the online self is that, if the Extended Mind Thesis and the Externalist View of personal identity are tenable, then the online self could be regarded as an extension of the original, offline self, which would mean that the online self is part and parcel of the offline self from the beginning. Furthermore, the analysis that has been offered here with regards to the offline self could in principle be applied to the online self also. In other words, the problem of how to account for personal identity and personal ontology can be applied to the online self as well as its offline counterpart. We could inquire what the constitutive elements of the identity of an online self are that appears in social networking sites such as Facebook and Twitter. And I would like to propose that the same basic elements that are used in analyzing the offline self can be also used with the online self also. Thus, the online self can be analyzed with regards to the problem of personal identity: What accounts for the identity of an online self? And here I propose that the same accounts that are tenable with the offline self are also tenable for their online counterpart too. This means that the Extended Mind Thesis and the Externalist Account of personal identity can also be used, profitably I believe, in analyzing the online self too.

This is only possible if the offline self and its online counterpart could be regarded as continuous with each other. If we follow the argument on the self as informational that I proposed earlier, then a way can be found to bridge the two kinds of self together and see how they are actually continuous. If the self is ultimately speaking informational, then since the online self is obviously informational (after all if the offline self is informational then the online self can be nothing else but informational too), then any kind of analysis that works with the offline self should work with the online self too.

However, there seems one clear difference between the two. The offline self can be a source of intention and action and subjective thought, and it is at least unclear how the online self can be the same. This is known as the Problem of Online Agency. If the online self is indeed continuous with the offline one, then does the former have the capability of subjective thinking and initiating action on its own? If an action is done by an online self, should the offline self that acts behind the online self be responsible? These are important problems that we will come back again and again throughout this book. However, since the problem deals more with ethical issues, it will be discussed more extensively in the next chapter. Here an intuitive answer will be given, which will be argued for in detail later. The continuity between the offline and online selves does not mean that the online self alone has the ability to think or to plan or have intention. After all the online self is only a phenomenon on the computer screen having no brain and no body of its own. However, since the online self is always a projection of someone who presents his or her own persona to the outside world through the means afforded by the information technology, the continuity is there even though sometimes it is difficult to trace who actually is the person behind this or that online self. This situation is not unlike that of a criminal who wears a mask or some other means of disguising his own identity when he commits the crime. Although he is hidden behind the mask, this does not mean that the mask itself is the one who does the crime. Hence it is not the online self per se that does the crime when there is an incident of computer crime perpetrated by the online self in question. Some real person has to be behind the act and it is the duty of the law enforcement officer to trace who that real person is. This perhaps makes a case for a law that requires Internet users to declare their own real identity when connected to the Internet, though certainly we have to be on guard against any possible violation of individual privacy (This issue will be discussed later in the book). However, having said this, I also would like to maintain that in some contexts it might be more expedient to talk about an online self as though he *himself* had the ability to think and act. This is akin to the situation where a masked man presents himself and is referred to as, say, "the masked man." Here the definite description becomes functionally a proper noun that when used in relevant contexts succeeds in referring to the man himself, not as his real identity (complete with real name, etc.) but in his capacity as the masked man to did such and such a thing. In the same vein, an online self presents himself in the cyberworld (which is continuous to the real world anyway), we may not know the real identity of the person behind the online persona, but we can certainly refer to him as if *this* online self were capable of thought, planning, intending, and so on.

Tying up the discussion on continuity with what we have seen in this chapter, we find that the problem of agency is directly linked to this problem because in order to claim agency of action performed by an online self, usually the person who stays behind running the online self or persona must be responsible. However the two are quite distinct. The problem of agency deals with who is ultimately responsible for an action performed in the online world by some online self, and the problem of continuity asks what the links are between an online self and her offline double. Commonsensically there has to be a continuity before there can be a claim of agency. In normal cases an offline user sets up an online persona or self and the relation between the two is much like that between an actor and his character. The actor can certainly play himself, but he can be many other characters who may have very different personalities from that of the user himself. If it is possible for an offline user to set up an online character whose personality is totally different from who he is in real life, then the continuity between the two appears to be severed. We can look at the issue in two ways. On the one hand, the offline user certainly controls the online persona; even though the personalities of the two are totally different, the former still controls the latter as an actor controls his character in a mask play. Furthermore it is also possible for the offline user to play many roles which are different from each other. On the other hand, as the personalities of the character and the actual user can be totally different, if the user is anonymous, then there is no way to trace the identity of the latter through analysis of the personality of the former. In the online world analysis of personality can be done by collecting the postings, comments, shared links, taggings, etc. that an online self leaves in the social network and try to match these postings with a real person in the offline world. The problem of continuity is concerned with whether it is actually possible to set up a new personality such that no one can trace that back to the real user. For the sake of brevity, let us call the thesis that there is a radical discontinuity between the online and offline selves as the Radical Discontinuity Thesis. Basically the Thesis says that there can be a situation where an online self is totally discontinuous from their offline user, so much so that they become totally different persons (in as much as one can coherently talk about persons in the online context). As we shall see, I argue here for a converse of this thesis—namely that there is always a bridge toward the online and offline selves whenever there is a connection between the two (in most cases the latter controlling the former), which we might call the No Radical Discontinuity Thesis.

It would seem on the surface that such radical discontinuity is possible. My concern is, strictly speaking, with the question whether an online identity could be radically different from her offline counterpart. The problem is very similar to what Dave Ward calls the 'Multiplicity Thesis,' namely the thesis that it is possible that there can be multiple online identities anchored by only one offline user (Ward 2011). After all, we are familiar with cases where a user assumes various guises in the online world, and we are also familiar with cases where a male user assumes an identity of a female online and vice versa. As the role playing and acting analogy shows, an actor can play several distinct parts, so too can the offline user, who can conceivably play many parts, none of which has any links back to original user. However, my belief tends to be that radical discontinuity can become increasing difficult to maintain in the more saturated online world where information is always increasing. In order to sustain radical discontinuity one has to maintain two separate sets of information, one about oneself in the offline world and the other in the online one. Suppose that the online world is so saturated with information that it becomes indistinguishable from its offline counterpart. Then this would mean that the person has to maintain two full distinct identities, always having to keep them separate all the time lest somebody get to know the secret. One is reminded of the case of Superman and Clark Kent; sooner or later the truth will come out, as we have seen in the movie when Clark's girlfriend soon knows who he really is after only a brief moment. Information tends to spill across boundaries, and in practice information kept in one circle (say those about Superman) almost has a way of leaking out to other circles (to those who know Clark Kent) even with the best intention to keep them apart.

In more mundane setting, we can imagine an average offline person who assumes an identity of, say, Superman, when he engages with his less powerful peer in the world of social network. Facebook has had a long standing policy of not allowing untraceable pseudonyms or anonymity in their network, but such policy is being constantly challenged, and Facebook is now seriously considering revising it if they have not done so already (Stone and Frier 2014). My argument is that it is increasing difficult to maintain in the real setting such two strictly separate identities. The user who becomes Superman in the online world cannot talk as if he were Superman in the movie with his friends all the time, because if he were to maintain sane conversation with his friends then he cannot forever talk about Kryptonite or of being born on an alien planet and so on, since his friends in most cases of real friends who inhabit the real, offline world that we all have direct access to, interacting with these friends require "Superman" to reveal his offline identity. That is almost a necessary condition for engaging with others in the social network. He can put up the image of Superman as his profile picture and writes in his "About Me" section that he was born on the alien planet, having been put in a spaceship by his parents to flee the exploding planet, and so on. But once he engages with his friends, he has to reveal who he is. If his friends are close enough, some bit of information from him is enough to let them know who he really is. In order to maintain strictly that he is in fact the "real" Superman, he has essentially to make new friends all the time in the social network because Superman of course never interacts with anybody in real life (because he exists only in cartoon books and movies). And those friends have to accept being befriended by "Superman" who always insists that he is the real, flesh and blood Superman from the planet Krypton. I don't think many would consider doing that. The upshot, then, is that Superman cannot in normal settings escape being eventually discovered to be Clark Kent, hence a refutation of the Radical Discontinuity Thesis. This does not necessarily imply that Ward's Multiplicity Thesis must also be denied because it might be possible that an offline user can maintain a number of online personalities, all of which are somehow continuous with her but are not exactly the same or similar with one another. As Schechtman argues, the continuity between the online self and the offline user can be maintained in a broader context of an account of a person. A typical husband and father may entertain a different set of rules and expectations when he travels to Las Vegas to unwind, but this does not necessarily imply that he becomes a radically different person, and so it is for the online context (Schechtman 2012, pp. 341–343).

In another vein, people are also usually attached to their avatar. Raffaele Rodogno also talks about identity through attachment, which underpins the feeling that one is identified with one's avatar (Rodogno 2012). Although their avatars do not necessarily resemble the real personality of the person behind, people are still attached to them because they feel that they own them and in a sense the avatars become an extension of themselves. Jessica Wolfendale argues that avatar attachment is an important aspect of someone's personality, so much so that when her avatar is harmed that harm extends to herself too. She writes: "Avatars are … far more than mere online objects manipulated by a user. They are the embodied conception of the participant's self through which she communicates with others in the community" (Wolfendale 2007). She cites a fairly large number of empirical studies that support the conclusion that people are emotionally attached to their avatars. However, the moral argument she is making is that avatar attachment is a morally significant concept in that when the avatar is harmed that action is unethical. She reasons that in as much as someone is harmed when her possessions are stolen or willfully damaged, someone is also harmed when her avatar is wronged. This is so because the avatar functions much like the possessions. One naturally feels emotionally attached to one's possession, and this is also the case with the avatar. Wolfendale's argument thus affirms the view that there is a continuity between the online and offline selves, even though the two may be very different from each other. Furthermore, attachment and the feeling of being in possession of the avatar are still there even when the avatar is an anonymous one, because the personality or the character of the avatar does not have to match that of the offline person for the latter to feel affinity toward it. Thus one can be attached to more than one avatar, which does not have to be exactly similar to oneself in real life. Here the Multiplicity Thesis is not refuted and the Radical Discontinuity Thesis is again denied. According to Rodogno, "[t]hough it will not be possible to formulate a unitary conclusion, we shall find a trend to the effect that it is, in most cases, misguided to think that the same individual may have distinct online and offline identities" (Rodogno 2012, p. 322). The identities can in fact be distinct, but not so much as to become totally different personalities with totally different backgrounds and histories; sooner or later these conflicting information will come out in the open and the cover will be blown, so to speak.

My critique of the Radical Discontinuity Thesis might raise a question whether how it is related with my earlier argument concerning the externalist conception of personal identity. After all, if the criteria used to maintain personal identity is external to that person's consciousness and belief, then the critique of the Radical Discontinuity Thesis might on the surface conflict with it because the critique seems to imply that the online and offline persons need to be somehow one and the same, whereas the externalist conception might seem to imply, on the other hand, that the two can be different (especially when the external condition is such that they can indeed be different). In other words, the externalist conception might seem to allow for a possibility of a radical discontinuity between the online and offline selves, which my argument here tries to deny. However, the externalist conception is in fact neutral toward the question whether the online self and her offline user must be one and the same or not. Thus by itself it does not permit any implication that the online and offline selves must be either identical or not. It is only an account of what conditions need to obtain in case the two persons are one and the same already. The Radical Discontinuity Thesis, on the other hand, says that it is always possible for the online and offline selves to be so different as to be two different persons. Hence an account of personal identity such as the externalist conception does not apply. The two principles are logically independent from each other.

4.10 Agency, Continuity and Philosophy of Technology

Through discussing the thoughts of some leading philosophers of technology as well as the two main philosophical problems surrounding the online self, I intend to situate the online self in the context of philosophy of technology. We have seen that philosophers such as Heidegger, Marcuse, Borgmann and Dreyfus tend to have a critical and quite negative view toward technology in general, which leads us to assume that they might not have quite a positive view regarding the online self either. Andrew Feenberg appears to have a more positive attitude, though his concern is mainly on the question of control and how technology needs to be considered as a part of a larger social and economic context. Viewed from the perspectives of the critical minded philosophers, the online self appears to become a symptom of the control that technology is having upon us. Both philosophers from the left and the right, such as Heidegger and Marcuse, tend to view the online self as something one needs to take a critical stance toward. Even though these philosophers of course were already dead long before the online self came to the scene, we can use their analyses of technology to gauge what they would have said were they around to experience it. For Heidegger, the online self would be a clear example of technology's enframing our lives. The appearance of the online self on the rectangular monitor exemplifies perfectly the frames through which we all are being subjected to, as if our own *beings* are there only within those frames. For Marcuse, the online self can be a manifestation of the strong power of the capitalist system that grips us in such a way that we do not even know that we are being controlled so tightly. The modern flat LCD monitor currently in vogue also fits Marcuse's analysis of the one-dimensional man. Just as man has become one-dimensional in the advanced capitalist society, so too is the online self, and much more radically, as it fits only within the flat screen. It is as if our lives are being reduced to no more than two dimensions on the flat screen. Here Marcuse and Heidegger curiously agree, although they come from different polar opposites along the political spectrum.

So the question is whether the online self is really a front that thinly veils the domination of technology over us, or whether it can reflect some degree of freedom or even a way toward resistance against overarching power. Before we deal with this very important question, let us review the necessary groundwork that has already been laid in the pages above. To the extent that freedom is possible for the actual user who has an online profile, she exercises that freedom in the cyberworld through the online self. Her interaction with others through the self gives an illusion that it is the online self itself that does the interaction. And this is not a bad illusion for it helps us understanding the pervasiveness of the online world and the destruction of any boundary that might exist between the online and offline worlds. The problem of continuity is also relevant as a means by which the two selves are interconnected. I have argued for the No Radical Discontinuity Thesis, namely that there can always be a way of linking the online and offline selves. When the self is constituted by information as I argued in the last chapter, eventually the two selves will come to share the same set of information so that they come to be recognized as one and the same. To keep them apart forever no matter how much information is there would mean that the two are always distinct from each other and no amount of information can reveal the underlying link. But this would mean, in effect, that the two are *really* separate because the underlying link will forever be unknown.

So what about the status of the online self in terms of the critical perspective of philosophy of technology? We have seen philosophers such as Heidegger and Dreyfus taking a critical and quite negative attitude toward technology, which leads us to conclude that they would not have endorsed the online self and its technological apparatus had they known about it and analyzed it. Furthermore, contemporary analysts such as Sherry Turkle, who talks about people today living "alone together" (Turkle 2011), seems to be critical in the same way. Though Turkle by no means says that we should throw our smart phones and tablets away, her analysis points to a contemporary phenomenon that cannot be ignored. People indeed appear to living alone together, each staying in close proximity with one another but is not interacting physically with anybody; instead they interact very much with their peer in the online world. It is almost as if the physical self has become a conduit for the online self to interact. Is that a good or a bad thing? My tentative answer at this point is that, like the same question in other areas, it depends. And as in other areas of life, too much of one thing cannot be good. Being totally immersed with the online world so that the physical world is all excluded does not seem to be a good thing; in the same vein, being totally immersed in the physical world with no online connection at all–in the environment where online connection is possible and has become another aspect of contemporary life–probably is not a good thing either. Furthermore, in the near future where the online and offline worlds are merging with each other so much that it is difficult to tell one apart from the other, the question does not even make sense. Technological determinists might object that this scenario only reflects the situation where technology completely dominates our lives; talking about the "near future" situation presupposes that the path of history has to be a one way street leading to such a future, and that cannot be a good thing. However, the near future I am talking about here does not have to happen. Things could happen otherwise that do not lead to this situation occurring at all. Alternatively, humans may get together and decide that the path that relies on sophisticated technologies is not the way to go and collectively decide to pull back and live as if they were in the seventeenth century. They could perhaps conceivably do that, but the chance of that actually taking place is quite small. The technological determinists would like to argue that technology travels in only one path and we are powerless to alter it. Thus when they see that the path of current development is such that it will lead to the situation where the online and offline worlds are merging they take that as an example for their critical outlook. There is of course nothing wrong with the critical outlook, but the assumption of the determinists that technology travels only in one straight path has been amply shown to be wrong in numerous empirical studies, some of which are discussed in Feenberg's work.

Nonetheless, I think technological determinism still has a point in reminding us of the potential danger that unbridled technological development could bring. As Feenberg and Ihde make clear, instead of passively waiting for things to happen, we can take things into our own hands and become more active in channeling the course that technology takes. If we don't like living alone together and if we still do not want to go back wholesale to the seventeenth century, we have to find a way to enjoy the benefits of technology (those things that we like) while avoiding the pitfalls (those we don't). Before doing that we need to get together and deliberate among ourselves what we, collectively, would like to have and what not. Here philosophers of technology can be helpful in reminding us of the need to ask tough, fundamental questions the way Socrates did in Athens. These questions, furthermore, can be a positive asset when future technology is designed too.

References

Allen, C. (2010). Artificial life, artificial agents, virtual realities: Technologies of autonomous agency. In L. Floridi (Ed.), *The Cambridge handbook in information and computer ethics* (pp. 219–232). Cambridge/New York: Cambridge University Press.

Bakardjieva, M. (2009). Subactivism: Lifeworld and politics in the age of the internet. *The Information Society, 25*(2), 91–104.

Bijker, W. E. (1995). *Of bicycles, bakelites, and bulbs: Toward a theory of sociotechnical change.* Cambridge, MA: MIT Press.

Borgmann, A. (1984). *Technology and the character of contemporary life: A philosophical inquiry.* Chicago: University of Chicago Press.

Borgmann, A. (1999). *Holding on to reality: The nature of information at the turn of the millennium.* Chicago: University of Chicago Press.

Borgmann, A. (2013). So who am I really?: Personal identity in the age of the Internet. *AI & Society, 28*, 15–20.

de Spinoza, B. (1985). *The collected works of Spinoza, Volume I.* (E. Curley, Ed. and Trans.). Princeton: Princeton University Press.

Dreyfus, H. (1978). *What computers can't do: The limits of artificial intelligence.* London: HarperCollins.

Dreyfus, H. (1999). Anonymity versus commitment: The dangers of education on the Internet. *Ethics and Information Technology, 1*, 15–21.

Feenberg, A. (1999). *Questioning technology.* London: Routledge.

Feenberg, A., & Friesen, N. (Eds.). (2012). *(Re)Inventing the Internet: Critical case studies.* Rotterdam: Sense Publisher.

Floridi, L. (2010). Information ethics. In L. Floridi (Ed.), *The Cambridge Handbook in Information and Computer Ethics* (pp. 77–100). Cambridge/New York: Cambridge University Press.

Floridi, L. (2013). Distributed morality in an information society. *Science and Engineering Ethics, 19*, 727–743.

Heidegger, M. (1977). The question concerning technology. In *Basic writings* (pp. 283–318). New York: HarperCollins.

Heidegger, M. (1992). *Parmenides (1942–43).* (A. Schuwer and R. Rojcewicz, Trans.). Bloomington: Indiana University Press.

Hongladarom, S. (2002). The web of time and the dilemma of globalization. *The Information Society, 18*, 241–249.

Hongladarom, S. (2008). Floridi and Spinoza on global information ethics. *Ethics and Information Technology, 10*, 175–187.

Hongladarom, S., & Kelly, M. (2004). Time, technology and globalization. *Journal of Philosophy in the Contemporary World, 11*(2), 55–62.

Ihde, D. (1993). *Postphenomenology: Essays in postmodern context.* Evanston: Northwestern University Press.

Ihde, D. (2002). *Bodies in technology.* Minneapolis: University of Minnesota Press.

Ihde, D. (2004). What globalization do we want? In D. Tabachnik & T. Koivukoski (Eds.), *Globalization, technology and philosophy* (pp. 75–91). Albany: SUNY Press.

Ihde, D. (2009). *Technoscience and postphenomenology: Peking lectures.* Albany: SUNY Press.

Ihde, D. (2010). *Heidegger's technologies: Postphenomenological perspectives.* New York: Fordham University Press.

Marcuse, H. (1991). *One-dimensional man: Studies in the ideology of advanced industrial society.* London: Routledge.

Rodogno, R. (2012). Personal identity online. *Philosophy of Technology, 25*, 309–328.

Schechtman, M. (2012). The story of my (second) life: Virtual worlds and narrative identity. *Philosophy of Technology, 25*, 329–343.

Stone, B., & Frier, S. (2014). Facebook turns 10: The Mark Zuckerberg interview. Bloomberg Businessweek Technology. Retrieved from http://www.businessweek.com/articles/2014-01-30/facebook-turns-10-the-mark-zuckerberg-interview#p3

Turkle, S. (2011). *Alone together: Why we expect more from technology and less from each other.* New York: Basic Books.

Ward, D. (2011). Personal identity, agency and the multiplicity thesis. *Minds and Machines, 21*, 497–515.

Wiegel, V. (2010). The ethics of IT-artifacts. In L. Floridi (Ed.), *The Cambridge handbook in information and computer ethics* (pp. 201–218). Cambridge/New York: Cambridge University Press.

Wolfendale, J. (2007). My avatar, my self: Virtual harm and attachment. *Ethics and Information Technology, 9*, 111–119.

Chapter 5
Selves, Friends and Identities in Social Media

This chapter investigates the phenomenon of online friendship. This is a very interesting phenomenon because with the advent of social media the role of "friends" has become much more visible. Social networking websites such as Facebook originated as an attempt to link friends together who already know each other in the life outside. Facebook, as is well known, originated in the dorm of Harvard College when Mark Zuckerberg and his friends wrote up the code of the website in an attempt to link their classmates together. Thus the website functions as a "Facebook" that is a real volume containing faces of the classmates and their information. One thing that separates the typical college year book and the new website is that users, that is, the friends linked by the site, can post information about themselves and others so that others in the group can see. This information posted by the users here then becomes the lifeblood of Facebook in the years to come (Stone and Frier 2014). The information is posted on a "news feed," a series of information that is constantly updated on an individual's Facebook page showing what is happening and who is doing what; thus the users are constantly provided with up to the minute update of what is going on their friends. So the information and the connection among people are the two defining features of social networking sites. From the perspective of an individual user, the information that is posted on her timeline always come from her friends, those that she know; hence Facebook appears to reinforce the ties that already exist among herself and her social group. We can thus say that the information is secondary to the users themselves. In other words the information coming from her social peer can be regarded as an extension of the identities of each of the member of her social group themselves, an idea that reflects the view that the self is constituted through information, a topic of our discussion back in Chap. 3. In this chapter, then, I would like to investigate more closely these "friends" that form the social group that an individual finds herself in. The main question is: What kind of friendship is there in the social networking sites? Can insights obtained from both Eastern and Western philosophies shed any light on how we should understand friendship as an online phenomenon? Suppose that we could talk about the relationships between online persons as one obtained among

© Springer International Publishing Switzerland 2016
S. Hongladarom, *The Online Self*, Philosophy of Engineering and Technology 25,
DOI 10.1007/978-3-319-39075-8_5

friends, what is the quality of online friendship when compared with the traditional one? In order to answer these questions, we begin by looking at what ancient and modern western philosophies have to say on the issue. Furthermore, we will also look at some strands of Eastern thought on their attitude toward friends and friendship too. Afterwards we will rely on these insights in our analysis of online friendship? Is online friendship a spurious phenomenon that no one can be a real friend with any one online? Is there actually such a thing as online friendship? Does becoming a friend require that one has to have a physical body? We will investigate these vexing questions in order to find preliminary sketches of an answer to them, and we start by looking back at some of the important strands in both Western and Eastern thoughts on friendship.

5.1   Aristotelian Conception of Friendship

For Aristotle friends are important as a means by which one achieves happiness, for one could hardly be said to be happy if one is solitary. In Book One of the *Nicomachean Ethics,* Aristotle writes:

> Still, happiness, as we have said, needs external goods as well. For it is impossible or at least not easy to perform noble actions if one lacks the wherewithal. Many actions can only be performed with the help of instruments, as it were: friends, wealth, and political power. And there are some external goods the absence of which spoils supreme happiness, e.g., good birth, good children, and beauty: for a man who is very ugly in appearance or ill born or who lives all by himself and has no children cannot be classified as altogether happy; even less happy perhaps is a man whose children and friends are worthless, or one who has lost good children and friends through death (Aristotle 1962, 1099a30–b10).

One sees here the value of friendship for happiness. Friendship is one of the "wherewithals" without which it is very difficult to achieve happiness. Happiness, or *eudaimonia*, is for Aristotle not the kind of static characteristic that describes merely a state of mind, but the term describes one who is engaged in activity, one whose nature is complete with human reasons and emotions that altogether make up for a fully virtuous person. Friends are important for a happy and virtuous person because without friends it is difficult to perform the kinds of action that are conducive to realizing the virtues. One who lives all by himself, according to Aristotle in the quotation above, cannot be regarded as happy. Perhaps certain virtuous action cannot be accomplished without the help of one's friends. In any case Aristotle pays a very strong attention to friends and friendship as a means toward supreme happiness.

In Books Eight and Nine of the *Nicomachean Ethics,* then, Aristotle discusses friendship in detail. For him friends are "most indispensable for life" (1155a5). It is an intrinsic good which provides meaning for all other goods. Rich men, says Aristotle, would find that their wealth would amount to nothing if the wealth is not used for "best works," which are done for the sake of one's friends (1155a8–9). Imagine a rich man who lives alone without any companion or friends, what

Aristotle seems to say here is that the rich man here would find his life to be devoid of meaning and significance because there is no one for whom the wealth and its usefulness is directed. The wealth cannot be directed toward the use of the rich man himself because that would incur no benefits to the community, which is necessary for any kind of good work. Furthermore, Aristotle asks how prosperity can be safeguarded without friends, implying that friends are needed even to keep the wealth. One cannot remain utterly alone when one is in possession of great wealth because that wealth needs to be maintained and a large amount of trust is required for the wealth to function properly. In short, without friends whom one can trust, the wealth of the rich man becomes almost nothing. In addition, poor people also need friends because friends help one another in times of need. And when one travels abroad one also needs to good will of one's friends who are one's hosts and depend on them for one's very own survival (1155a10–22).

For Aristotle, genuine friendship occurs when one wishes for the good of one's friend for his or her own sake: "Those who wish for their friends' good for their friends' sake are friends in the truest sense, since their attitude is determined by what their friends are and not by incidental considerations" (1156b10–15). Having discussed three kinds of friendship, namely one based on the motives of what is good, pleasant and useful (1155b20), Aristotle argues that the best form of friendship is one where the motive of becoming friend is not just to find benefits for one's own sake only, but for the sake of the friend and the friend must also recognize this same motive too. So the friendship is mutual and reciprocal: One wishes for the good for the sake of one's own friend and the friend also recognizes the good in oneself for the sake of oneself, who is his friend, also. For Aristotle, then, "to be friends men must have good will for one another, must each wish for the good of the other on the basis of one of the three motives mentioned, and must each be aware of one another's good will" (1156a5). The mutual recognition of the good in the other for the other's sake is what makes for a perfect friendship: "The perfect form of friendship is that between good men who are alike in excellence or virtue. For these friends wish alike for one another's good because they are good men, and they are good *per se*, (that is, their goodness is something intrinsic, not incidental)" (1156b5–10). This perfect form of friendship obtains when both recognize that the other has moral excellence and thus wishes that the other achieve blessedness and happiness without thinking of merely enriching or finding pleasure to oneself alone. Since friendship must be equal, mutual and reciprocal to be genuine, the wish for happiness in one's friend must also be there in both friends who wish for the other. Thus Aristotle says that his kind of friendship is there only among good men; that is, only among those who have attained a high degree of moral excellence. In other words, the perfect form of friendship for Aristotle is a reciprocal kind where the relationship is between equals. The friends have to be equals—they must be alike in terms of moral attainment and in other areas too. Aristotle also says (1158b10–25) that if the statuses of those who are friends are not equal, such as those of father and son, or husband and wife, then the form of friendship is not perfect, but an imperfect one. Father and son can still be genuine friends, but the relationship is not symmetrical. The son owes something to his father in a way that is different from what the father owes to the son. The son may have to respect his father and acts as a good son, but the father cannot act as a good son, because he is the father and has to act toward the son as a good father. This is not the case with equal friends whose quality and content of the act toward each other is the same.

Aristotle maintains that friendship is a means toward happiness, and the ideal of character friendship is that of two people who are alike in virtue and wish for the good of the other for their sake (1157b–1158a). One cannot achieve happiness by oneself alone, since having friends who share in the good things in life is essential, as we have seen. Thus Nancy Sherman views the situation as one in which for Aristotle friendship is an extension of the self, in other words, the friend is "another self," a mirror image of a person, which nonetheless is compatible with the idea that each person is his or her own individual (Sherman 1987). It seems that character friends must be so alike to each other that they could be mirror images of each other, since they have to share equally in all kinds of virtue and be equally virtuous. However, Sherman sees that these equally virtuous persons need not be exactly the same, as one may have more virtue in one area than the other, while having equal amount of virtue over all (Sherman 1987, p. 609). One friend may have more virtue of generosity, while the other is more pronounced in his virtue of courage; however, the two friends here are character friends in that they both have all the virtues, only that one has more courage and the other more generosity. In other words, the two friends are both alike and different, which creates a dynamic by means of which friends can both look at the other as an extension of her own self and at the same time gains more self-knowledge. One can only increase one's self-knowledge through observing the other who is not completely alike oneself, but who is not so different as to find nothing that is applicable. Here one finds an interesting connection between friendship and the self. One's friend according to Aristotle as an extension of one's own self. It is in other words one's second self; however, this does not mean that the friend is one's mirror image, but someone who is sufficiently alike and sufficient different that makes an awareness of aspects of one's own self possible which would not have been had there been only oneself alone without the friend. This account of friendship, then, fits rather nicely with our previous discussion of the extension of the self and externalism regarding personal identity. We have seen in Chap. 3 that the boundary of the self is not necessarily limited to the skin, and as our friend is in a way an extension of our own self, then we find a fusion of two selves that make it apparently the case that selves are not limited to the skin only. According to Aristotle the achievement of the supreme goal of living, attaining *eudaimonia*, is not possible without someone to share the happiness with and someone who is actually necessary for the achievement of the happiness in the first place. And as we have seen in previous chapters that the online situation is only a novel way in which the self can be extended, then it looks increasingly likely that online friendship in itself is not inimical to the kind of friendship that makes achievement of happiness possible. It is clear in any case that online friends are not the only way, but at least they could be sufficient in their own way, which will be specified later in this chapter. It does not look too far-fetched to say that one's genuine friend (in Aristotle's sense), being an extension of oneself, is one's own self extended out to another person. The extension is not complete; otherwise the friend would be just our perfect clone, but the friend has some elements of what make up myself, and I also have some elements that make up the identity and the self of the friend too. This does not destroy the identity or personality of either of us because we still retain our respective uniqueness through the information that makes up who we are individually. It is only that some elements of what make up who both of us are shared by us, while the composition that makes up who each of us is unique to each of us. We will come back to these topics more fully in the section on the analysis of online friendship later in the chapter.

A series of questions arise naturally when we look at the phenomenon of online friendship in social networking sites with an eye toward Aristotle's view on friendship is how one could characterize online friendship in Aristotelian terms. We have seen that Aristotle places a very important place for friendship: One could not achieve supreme happiness, or one would find it extremely difficult to do so, without having friends. But what kind of friends are they? Do the online friends suffice? Can online friends be included as instances of the description of genuine friends or the highest form of friendship that Aristotle discusses? Suppose one does not have any offline, physical friends but a large number of online friends, can one still achieve *eudaimonia*? The answer I am giving in this chapter is that there is nothing in online communication that would, in principle, prevent some friends from attaining *eudaimonia*. Moreover, the highest form of friendship in Aristotle's sense is also attainable through online means. We will try to unpack the arguments for these assertions as we go along in the chapter. Basically the argument is that characteristics of offline friends, having flesh and blood, physical presence and the like, do not seem to be entirely necessary as qualities of genuine friends. In other words, it is possible that one can have virtuous relationships with one's online friends in the way that Aristotle would approve. The mutual recognition and reciprocal wish for the good for the other for their own sake can, it seems, be accomplished with online friends as well as offline ones.

In Book Eight Aristotle analyzes friendship as both a characteristic and an activity, and he also claims that it is not possible for one to have more than a handful of real friends. In Section 5 of Book Eight Aristotle says that friendship needs to be maintained by constant activities among friends: "Out of sight, out of mind," says he at 1157b10. However, a note by the translator indicates that the original Greek literally translated is "A lack of converse spells the end of friendship" (Aristotle 1962, p. 223). The original Greek is "πολλὰς δὴ φιλίας ἀπροσηγορία διέλυσεν," literally "Many friendships become parted asunder because of lack of discourse." Friends need to maintain their activities together and these activities are constituted mostly by conversation, as we see in the literal translation; otherwise the quality and even the existence of the friendship itself may suffer. In this case a possible interpretation might be that the 'discourse' that friends need to engage with one another necessarily consist of shared activities that they do with one another. It is not enough, so goes the interpretation, that friends merely converse with one another to maintain their friendship. However, this is not supported by the text. The word ἀπροσηγορία means 'want of discourse;' if Aristotle wants to emphasize the meaning that the discourse here includes doing something together in addition to conversing, he should have said it directly, using another word. Another reason in support of this is that engaging in shared activities is an important Aristotelian concept—this is after all what members of a polis do in order to realize their basic identity as full individuals. But here Aristotle appears to say only that it is *communication* that sustains friends together. In fact in Section 9 Aristotle discusses friendship among members of the same polis: "…at least men address as friends their fellow-voyagers and fellow soldiers, and so too those associated with them in any other kind of community" (First sentence of Book 9). If there is anything regarding the need for friends to engage in shared activities together, Aristotle should say it here. However, Aristotle goes on to say:

> And the extent of their association is the extent of their friendship, as it is the extent to which justice exists between them. And the proverb 'what friends have is common property' expresses the truth; for friendship depends on community. Now brothers and comrades have all things in common, but the others to whom we have referred have definite things in common-some more things, others fewer; for of friendships, too, some are more and others less truly friendships. (Book 9, beginning).

The picture of friendship that emerges here is that of friends to the extent that the members of the same community can be regarded as such; that is, to the extent that they engage in shared activities that are required for citizens to do in order to maintain the community and to participate in civic life. However, the extent of this friendship is the extent of their participation, which is defined also by justice. In other words, to the extent that these friends engage in civic life together, they cease to become friends once the bound of justice is broken among them. This is a different kind of friendship than the best kind that Aristotle talks about earlier. Actually this type of relationship should be understood more as comradeship rather than friendship, for in engaging in civic life and participating in it with one's fellow citizens, one does not need to know the other well enough to qualify as true friends in the sense discussed earlier. Furthermore, in the last sentence of Section 9 Aristotle says, "All the communities, then, seem to be parts of the political community; and the particular kinds friendship will correspond to the particular kinds of community." This shows that different kinds of friendship correspond to different kinds of community: soldiers, for example, have a kind of friendship that obtain among themselves, so do other groups. This is why people doing the same kind of things get together to form associations, and in the polis the citizens naturally get together to do things together that are the works of the polis itself. To this extent they are, in a sense, friends, but this kind of friendship is not the same as the truest kind which is defined by exchanging the selves as discussed earlier. It is indeed possible that in a very ideal situation all citizens might be truest friends with one another, always wishing one another all the best and so on, but Aristotle does not seem restrict the term 'friendship' to this ideal kind only. The upshot, then, is that shared activity and participation in a political community are not necessary conditions for this type of ideal friends, and are necessary only for this specific type of friendship that is defined by political membership or membership in specific groups.

One may also interpret Aristotle's important saying here as saying that friends need to be physically present close to each other. After all, the discourse that is required in maintaining friendship traditionally implies that friends need to be close to one another in order to talk with one another. In the same paragraph, Aristotle says "When friends live together, they enjoy each other's presence and provide each other's good. When, however, they are asleep or separated geographically, they do not actively engage in their friendship, but they are still characterized by an attitude which could express itself in active friendship" (1157b5–10). This could be interpreted as arguing that physical presence is necessary to maintain friendship; otherwise it is very likely that the enjoyment of each other's presence and the provision of each other's good will not be possible. Furthermore, when friends live far apart, then they do not engage in maintaining their relationship simply because communication is difficult. It is quite clear, then, that it is paucity of communication between friends that is at issue, and probably not mere geographical distance. In other words, it is lack of communication that is the culprit, and merely living far from one another. However, in today's world with the mobile phone and the Internet, communication is instantaneous and does not respect any distance. In this case friends can always maintain their communication link as much and as often as they like, so even when they live in opposite corners of the world they can still engage in conversation, which for Aristotle seems to be what is necessary to upkeep the friendship. If this is the case, then, it is possible for friends to remain friends with their activities maintained through online communication. Those who disagree, however, might counter that the quality of the activities that keep the friendship cannot be compared with face-to-face communication. That is certainly true, but since Aristotle argues that it is conversation that is necessary, geographical distance then might not have to result in loss of conversation as it certainly did in his time. Before the age of the telephone and the Internet, people communicated by writing letters when they were geographically separated. Friendship maintained through letters does not perhaps have the same quality as one with physical presence, but at least it is a kind of friendship and fits with Aristotle's own definition of wishing good for the other for their own sake. If this is the case for letter writing, then I don't see why this should not be able to carry over to the age of emails, Skype calls and instant messaging. Furthermore, as we have seen that shared community activity is not necessary for friendship, friends living far apart do not have to cease to be friends simply because their geographical distance makes it difficult to engage in political activities together. They can engage in joint online political activities, or they can help their local communities while maintaining their friendship through the online means. Most criticisms of online political activities seem to focus more on the quality of such activities than on the question whether engaging in online political activities is in itself insufficient for friendship. For example, the criticisms may focus on the existence of "flame wars" that happen often in online political communication, or "slacktivism" where one engage in political activity simply by sitting still in front of a computer only. Nonetheless, the criticisms focus more on the quality of political communication and activity sharing than on these acts themselves. In case where the communication and the sharing of activities is of good quality, such as when the online activity results in genuine deliberation of ideas that lead to concrete changes in policy that serve the goals of the community better, then we can see that it is the *quality* of the communication itself that matters, and not online communication *per se.*

The other issue that might be raised by those who do not trust online friendship is that Aristotle writes at 1158a10–20 that it is not possible for one to have too many friends:

> To be friends with many people, in the sense of perfect friendship, is impossible, just as it is impossible to be in love with many people at the same time. For love is like an extreme, and an extreme tends to be unique. It does not easily happen that one man finds many people very pleasing at the same time, nor perhaps does it easily happen that there are many people who are good. Also, one must have some experience of the other person and have come to be familiar with him, and that is the hardest thing of all. But it is possible to please many people on the basis of usefulness and pleasantness, since many have these qualities, and the services they have to offer do not take a long time (to recognize) (1158a10–20).

What Aristotle is driving at is that one cannot practically have too many genuine friends (where genuine friends are those that make perfect friendship in his sense possible). This is because the amount of investment of time and effort in loving someone, who is always a unique individual, is such that loving more than a few people at a time is not feasible. And this certainly goes with having genuine friends. Having a friend and maintaining the relationship with him or her requires time and effort, which makes it practically impossible to have more than a few genuine friends at a time. However, this is the case both in the offline and online worlds. It is certainly possible for one to maintain a small number of genuine online friends that satisfy all of Aristotle's requirements. Even in the online world, one can have friends of different characteristics and quality, that is, a smaller number of genuine friends with whom one interact frequently, and a larger number of more ordinary friends or acquaintances with whom one only occasionally interacts. Apart from the issue of whether online friends are, as online friends, of necessarily lower quality than offline ones (which we will take up in a moment), there does not seem to be any qualitative difference regarding the point that one cannot have too many genuine friends.

Another argument that could be levelled against the idea of online friends is that on social networking sites it is possible to have thousands of "friends" which is clearly not possible in the world outside; hence these "friends" are not real friends at all. These "friends," in other words, are not friends in the Aristotelian account because there cannot be any amount of investment of time and effort to retain any level of relationship with any of these thousand "friends." However, when one has more than a thousand "friends," chances are that these request the connection because they know the user and would like to get the latter's posts and comments on their timeline, just like a newspaper reader subscribes to a regular columnist in a newspaper. In this case the user with thousand friends becomes a provider of content, which is followed and taken up by the public. In response to this problem, Facebook has recently set up a "follow" feature where a user can follow another user without sending a request to become a friend. This feature has been with Twitter from the beginning. One follows another user on Twitter simply by clicking a button, and whatever posted publicly by that user will immediately appear on the follower's timeline. In imitating this Twitter feature, Facebook is acknowledging a change in their policy and their tradition. Instead of looking their site as a platform for real life friends who already know one another in offline life, Facebook eventually acknowledges that their platform is no longer one where those who know one another in real life interact, but people get to know and have connections with others whom they do not know at all outside of the Internet. On Twitter a user, especially well-known or "celebrity" personalities, can have followers in millions. It is thus not possible for the celebrity to maintain any level of acquaintance with the millions here. But are they the user's friends? It does not seem so on Twitter. On Facebook, however, one can distinguish between a follower and a friend. A friend is one who makes a request to the user and the user has to accept the request in order for the two to become friends. But to follow someone one just clicks the button without sending a request to "add" a friend, just like on Twitter.

One then wonders whether the "friends" on Facebook are really friends as traditionally understood in offline life. After Facebook has expanded to cover all corners of the globe, with more than five billion user base, it has tremendously expanded much more than what it was at the beginning as a platform for people who already know one another to get in touch. It is common for an average Facebook user to have as her "friend" someone she does not know in real life; moreover, in Thailand the practice of using pseudonyms is widespread, making it the case that a user might accept a friend request from someone whose real name she even does not know. Not only is a pseudonym or a totally invented name is used, the profile picture does not reveal the identity of the user who is behind either. In the Thai Facebook scenario, it is rather common for users to pose any kind of picture as their "portrait" in their profile. So it is not possible to know the identity of the real person who hides behind all this, and there is no way other users will know the real identity unless the anonymous user directly reveals his or her identity. Nonetheless, it is still possible for a kind of friendship to develop, because even though the real identity of the user is hidden, the user, under her new guise and new identity, does maintain a stable personality, so that she becomes a distinct person in the online world. When A gets to know B else in real, offline life, A will form a set of information that together works at identifying B out of the large number of people that A knows. These identifying characteristics work similarly to the definite descriptions in philosophy of language that successfully pick out a distinct individual object and none other. The definite descriptions thus function as a set of identifying information that help the user picks out a unique individual. It is through this mechanism that an online user successfully identifies another online individual and engages with her as a distinct individual in the online world. In offline life, when A becomes a friend of B, A has to be able to pick out B from among thousands of other people; otherwise we would say that A hardly knows B at all, and A cannot then claim to be B's friend if she does not know B. The fact that A knows B then depends on the set of uniquely identifiable information that serves to pick out B in A's mind. This is a necessary condition for A and B being friends (for of course B has to be able to pick out A in the same way too). This act of uniquely identifying someone to be such and such person that one knows does not seem to be limited to the offline life. We can imagine a case where in the online world A* and B* become friends on a social networking site; the two have not known each other before and it is possible that one of the two, or both, use pseudonyms so there is no way to trace the identity of either by linking the information obtained online with its offline counterpart. Here A* must be able uniquely to pick out B* from among the large number of his online friends that he is engaged with on the social networking platform, and B* must be able to do the same too. The situation parallels that in the offline world. So if A and B can become real friends in the offline world; so too can A* and B* in the online world. One might object that in the case of the offline world one has other clues besides what one can see on the monitor; for example in the offline world A and B can touch each other's body. But then one wonders whether the ability to touch someone's skin is a necessary condition for A and B becoming friends. If being a friend is as Aristotle says, that A and B need to be able to achieve something good for the other for the other's sake, then we have seen that this condition can indeed obtain in the online world. Hence there does not seem to be anything in principle against accepting that genuine friends can happen in the online world too. In short, it seems possible that online friends can satisfy Aristotle's requirements for genuine, virtuous friends (thus exemplifying what is called "character friendship"), those who recognize the good in the other and wish only for the good for the other's sake, there is nothing in the idea of online friendship that makes it necessary that this requirement cannot be satisfied.

There is an important caveat, however. By arguing that online friends could exemplify Aristotle's view of ideal friendship, I am not suggesting that all instances of online friendship are instances of the ideal. As it is not the case that any instance of offline friendship exemplifies the ideal of character, or virtuous friendship, online friendship should thus be considered likewise. What I am arguing is only that it is possible for an instance of online friendship to approach and eventually become the highest form of friendship. The online and offline worlds are fast becoming one and the same world, and this is the case also with friendship. There are a number of conditions that must obtain in order for online friendship to become a genuine, or character friendship. For example, the friends must be equally virtuous; they need to take the interests of the other into account and act for the other's sake, and so on. One should note moreover that Aristotle does not say that friends need to be embodied to exhibit the ideal form. Of course the Internet did not exist in Aristotle's time, but letter writing was available then and it is conceivable that two persons could become very good friends through correspondence. Perhaps both might have known each other before in real life and then maintained their friendship through correspondence. It is further imaginable that two persons could become friends who have not had a chance to meet in real life before, but there does not seem to be anything in Aristotle's account that friends need to be embodied to exhibit the ideal form. The main thing is that they wish for the good of the other for the latter's own sake, and this can well be obtained through correspondence, or with emails or social media in today's world. Certainly the quality of online friendship, in terms of immediacy and the ability to have face-to-face encounter and to perform activities together in the offline world, is missing, but this diminished quality of online friendship does not appear to detract necessarily from its possible status as an instance of the ideal form. On the contrary two people might be living very close to each other physically, perform activities together in the same place, etc., yet they are not friends at all.

5.2 Kierkegaard, Nietzsche and Online Friends

Apart from the Aristotelian conception, we should get a clearer understanding of the phenomenon of online friends and online selves through a look at What Kierkegaard and Nietzsche have to say on the topic of this chapter. Both have something very interesting to say about friendship, which supplements nicely Aristotle's view as we have just seen. The concern of this section is that there is an interesting angle in which Kierkegaard's and Nietzsche's view on friendship appear to illuminate our understanding of what online friends could actually be.

According to Kierkegaard, the main difference between the pagan and Christian conceptions of love is that the former does not know anything about universal love (Kierkegaard 1991). The notion of universal love, the love that knows of no particular personality or uniqueness of the object of love whatsoever, is completely alien to the pagan mindset. Erotic love and friendship, for Kierkegaard, must always have some specific object. Erotic love and friendship cannot just spread around and take whatever being as their object. Kierkegaard, on the other hand, says that if one can love any person whom one meet first when one walks out of the door without caring at all who that person could be, then one has the Christian conception of universal love (Kierkegaard 1991). It is a kind of love that knows no bound. The flipside, however, is that universal love and compassion is impersonal. One has this kind of love if the love must be equal to any being whatsoever without any prejudice or preference. In this sense, then, one has to love one's own mother and a complete stranger equally, and one's own action must correspond with the impartiality and equality of one's love. We can certainly criticize the Christian conception of universal love in this way, but fortunately this is not our concern in this chapter.

The main concern is to see how these conceptions fare when they come to today's world of online friendship. According to the teaching on universal love and compassion, any sentient being at all is an equal object of our love and compassion, and thus the being is our friend. Since the love and compassion is strictly impartial, all beings in the world are equally our own friends. Can this conception of universal friendship exist in the online world? One can certainly conceive of a scenario where one is impartial and totally objective in one's love and care toward everyone whom one knows or has contact with online. If I have no preferential treatment of all my Facebook friends in any way, then I might start to approach the ideal of universal friendship here. In fact I have to treat my online friends, those whose names are on my friends' list, and those who are not my friends equally. The beings outside and inside of the list have to be treated and regarded equally, so the whole idea of having a friends' list does not make any further sense in this conception. There does not seem to be anything in being online that would prevent one from being a universal friend of somebody else. In fact one who practices universal love here has to treat of the online friends and offline ones equally, and hence the boundary between the online and the offline breaks down.

Against this proposal some might object that such universal love is an ideal condition and challenge one to come up with a concrete example of someone who actually exemplifies true Christian universal love to all her friends in the online world. The presupposition is that it is very difficult if not impossible to find such universal love, thus the proposal is an overly ideal one which is not practicable. It is true, however, that true universal love is extremely rare; it is rare not only in online world, but in the offline world too. Nonetheless, that universal love is rare does not necessarily imply that it is not possible. In any case it exists as an ideal, a target for which one should aim for in order to perfect oneself. If the ideal is possible, then the point I have made is that universal love as discussed by Kierkegaard could actually exist in the online situation. There does not seem to be anything inherent to the online world or in online communication that prevents such love from being actualized.

However, this does not mean that online and offline friends are exactly the same in all aspects. It is clear that we can touch and see our offline friends face-to-face (if the friends happen to live close together physically) and thus close physical friends can do much more things together than friends who live far away from one another. The shared activity that Aristotle thought to be very important for virtuous friendship can be accomplished in many more ways and with more quality than the activity that online friends can do together (which is not much more than engaging in online conversation and sharing files with one another). All these have to be conceded. Nonetheless, the quality and the multi-dimensionality of offline friendship does not necessarily detract from the character of online friends if they are objects of universal love and compassion in the Buddhist and Christian sense, for to differentiate one's friends to be either offline or online would be to engage in preferential treatment which is inimical to the ideal of universal friendship.

While Kierkegaard emphasizes the friendly love of one's neighbor that does not distinguish one object of love from another, Nietzsche, according to Miner (2010), pays particular attention on the agonistic nature of friends as a main characteristic of genuine friendship and the shared goal of friendship which is truth. Nietzsche writes about this in the *Gay Science* (*GS*):

> Here and there on earth we may encounter a kind of continuation of love in which this possessive craving of two people for each other gives way to a new desire and lust for possession—a shared higher thirst for an ideal above them. But who knows such love? Who has experienced it? Its right name is friendship (*GS* 14) (Miner 2010, p. 57).

The "ideal above them," according to Nietzsche, is truth itself. Friends share the same passion together and the craving for each other in the normal sense is transcended by a "new kind" of desire, the desire for something that exists over and above both friends. In fact it is the mutual desire for truth that distinguishes a pair of friends from a pair of lovers. Truth is the ideal above the friends, and is something that binds them together. According to Miner:

> Nietzsche reveals its name when he envisages a fellowship consisting of himself and his close friends who "apply the standard of their criticism to everything and sacrifice themselves to Truth. What is bad and false has to be exposed!".... The ideal to which friends are devoted, the goal of their common voyage, is truth (Miner 2010, p. 57).

Nietzsche reveals the name of the "ideal above" the friends as truth. Friends "apply the standard of their criticism to everything," something that immediately reminds the reader of Nietzsche himself who applies his technique of genealogy and philological studies to morality and other phenomena. Friends "sacrifice themselves to Truth." Thus one can surmise that the preferred way of sharing activities among friends for Nietzsche is a common pursuit of truth. Nietzsche says "What is bad and false has to be exposed!" It is not enough merely to search for scientific, empirical truths which mundane and lack real significance. Nietzsche regards the common search for truth among friends as a moral activity: Friends perform moral acts together when they seek to gain access not only to empirical truths, but more importantly *moral* truths. These consist in the recognition and realization of what is "bad" and "false." Presumably these lie hidden from view and it requires a scholar of Nietzsche's acumen to see through the deception laid out to conceal these bad and false things. These may present themselves as "good" and "true" but for Nietzsche they are only a front that covers the fact that in truth they are exactly the opposite. Friends help each other out by opening up these concealed truths and they then reveal them for what they really are.

Furthermore, Miner also sees Nietzsche to be claiming that the most suitable relationship among those who are friends to each other is an agonistic one where the friends argue against each other. This agonistic nature, however, does not mean that friends have to be in conflict with each other, but they do challenge each other, keeping each other on edge. In this sense, friends cannot be fully like ourselves, and it is the agonistic nature of the friendship that is instrumental in uncovering the concealed truths, in finding out what is "good" or "bad." Here Nietzsche is far more sympathetic to Montaigne than to Cicero, whose view on friendship is that of two persons who are alike each other. Montaigne, on the contrary, writes: "I like a strong, manly fellowship and familiarity, a friendship that delights in the sharpness and vigor of its intercourse, as does love in bites and scratches that draw blood. It is not vigorous and generous enough if it is not quarrelsome, if it is civilized and artful, if it fears knocks and moves with constraint" (Montaigne 2003, p. 856, quoted in Miner 2010, p. 61). The idea is that Nietzsche prefers the friendship that "delights in the sharpness and vigor of its intercourse;" that is, friends that are not afraid to argue with their friends in order to search for what is in fact the case and what lies beneath the veil that needs to be uncovered. In order to search for the truth, friends need to argue with each other, perhaps sometimes they even have to endure "bites and scratches that draw blood." Nietzsche writes in a passage in Section 2 of *Schopenhauer as Educator* that he prefers Montaigne as one who writes "truly" and adds "to the joy of living on this earth" (Nietzsche 1990, quoted in Miner 2010, p. 61). This emphasis on friends arguing with each other aiming to uncover the truth, then, shows that for Nietzsche the overarching objective of friendship is a cognitive one. Friends get together in a common search for truth and they engage in arguments in order to achieve that. Truth is a cognitive concept, as it is expressed in linguistic form. Hence it lends itself naturally to the communication that friends naturally engage with each other. Arguments are verbal activities and truth is a linguistic concept. Both then lend themselves smoothly to friends existing in the online world. One can imagine friends who engage in verbal fights through emails or other forms of computer-mediated communication. Since the aim is to arrive at the truth, then it is a relatively easy matter to imagine the friends sharing this agonistic, truth-searching activity online. Since the aim is truth, what matters the most is the cognitive and semantic nature of communication. Here other things such as the physical body and location of the friends are not too important—which is the point I am stressing in this chapter: Online friends are not essentially different from offline friends and the noblest kind of friendship could also exist among online friends too. The "bites and scratches" that Montaigne talks about and that Miner takes to represent Nietzsche's view of truth-searching friendship are thus metaphorical. One bites and scratches the other verbally through the online medium as well as offline when one talks with the other face to face. This is not to deny that Nietzsche pays attention to the physical body; after all he talks quite a great deal about health and physical well-being, but in the context of friends who engage in the common task of searching for truth, health and physical well-being are not the primary goals. Or if they are they are taken more in metaphorical context, where one is "healthy" when one follows Nietzsche's recipe of the strong individuality and creativity, arriving ultimately at the ideal of the *Übermensch* or "overman" who, by the way, does not have to be actually physically strong and healthy at all. If we focus on this aspect of searching for truth through argument, then it is perhaps not too far-fetched to think that the *Übermensch* himself or herself could exist in the online medium too (For studies of Nietzsche and Buddhism, especially on the role of the ideal person, see Hongladarom 2011 and Panaïoti 2013).

5.3   Online Friendship and Authenticity

In their article on "Unreal Friends" (Cocking and Matthews 2000), Cocking and Matthews put forward an argument showing that it is not possible for one to have a genuine friend in an online situation. It must be noted that Cocking's and Matthews' article predated Facebook and other dominant social networking sites today, such as Twitter and Google+. What the authors have in mind in their analysis is thus the older form of web-based communication forum that developed out of the even older Bulletin Board Systems (BBS's). What is characteristic of this older form is that it is almost entirely text-based and does not have the "friending" feature that today's social networking sites have. The older social networking sites such as MySpace and Friendster were founded in 2003 and 2002 respectively [Wikipedia, http://en.wikipedia.org/wiki/Friendster and http://en.wikipedia.org/wiki/Myspace]. It remains to be seen whether Cocking and Matthews would revise their analysis in light of these newer forms of social media. Nonetheless, their analysis remains relevant especially their critique of text-based form of communication with remains, even on Facebook, the dominant form of interacting among friends. Friends in such situation, such as in social networking sites, are "unreal" in the sense that they are only called friends without having the characteristics that qualify them to be real ones. His argument is premised on the contention that the Internet is always limited in its dimensions, while the real, offline world offers more ways of looking at things in such a way that cannot be matched by the Internet. This limited dimensionality of the Internet then makes it impossible for friends to present themselves in all dimensions, as is dispositionally possible in the case of the offline world. Cocking and Matthews thus argue that the limited dimensionality of the Net then implies that on the Net things are always arranged and packaged, whereas in the real world things cannot be so packaged for presentation because of the myriad ways a thing can be presented and viewed. An online friend, as a consequence, thus presents himself in a way that he wants to be viewed by his peer, for he seems to be limited in how he can present himself on the Internet. This is contrasted with the offline situation where no matter how much he attempts to present himself and prevent others from observing the aspects of himself that he does not want to present, those aspects always have a way to be perceived by others. For Cocking and Matthews, then, friends become real in that way.

Perhaps the most that Cocking's and Matthews' argument can show is that friends on the Internet are always limited in how they present themselves, but then he needs further argument to show how being so limited necessarily implies that online friends are unreal. This is a crucial point. Friends, or things in general for that matter, can be "unreal" in several ways. One way is of course in the sense that Cocking and Matthews presuppose—online friends are unreal because they are, in his eyes, always the result of conscious and intended doctoring; for him online friends are always prepackaged and staged, just like actors are on stage. In this sense virtual friends are unreal just as actors are "unreal" on stage. When we see Julius Caesar on stage we are not of course seeing the Dictator himself there, but only someone else who is playing his character. So what we see is not the real Julius Caesar. Nevertheless, there is another sense of "unreal" where we say of illusory things as being unreal. The mirage of water on a hot asphalt road is unreal simply because there is no water there. Since online friends are limited in their dimensionality, they can be unreal in this sense. It appears, then, that Cocking are conflating these two senses, since he also makes a lot of use of the limited dimensionality of things on the Internet. Or perhaps Cocking purposely presupposes these two senses of "unreal" in his analysis. In such a case, then the link between the statements that the Internet is of limited dimensionality and that friends there are unreal can be understood as premised by this sense of "unreal" where whatever is of limited dimensionality is unreal because it is illusory, i.e., not representative of a real thing. This also is added the first sense of "unreal" mentioned earlier where the character Julius Caesar on stage is not the real Caesar.

However, it is not altogether clear that whatever is of limited dimensionality must be unreal. An image in a mirror is certainly unreal, qua the real thing purportedly represented by the image. We of course do not mistake an image of a pizza in a mirror to be the real pizza. But qua image we do not seem to have any qualm in maintaining that it can be as real as it gets. It is a real image. Thus when our friends present themselves in the online world, we do not thereby dump them unceremoniously as being unreal. In the case where those who present themselves online are those whom we already know in the non-virtual, offline world, then there is no reason for us to say that they are unreal simply because they appear on the Net. Furthermore, in the case where the friends are those we only know on the Net, there is no automatic reason we should classify them as unreal either. Perhaps there will be a chance for us to get to know these virtual-only friends offline, thereby linking the two worlds together. Or if we do not have a chance to meet in offline life, we might learn that our online friends have their lives in some far away, but real places such as another continent from the one we live, and so on. Here there is little reason to regard them as unreal either. Even when our friends present themselves in an environment that makes it necessary to limit the options of presenting and gaining access, such as on the Internet where limited dimensionality has been the norm, then there does not seem to be any reason why our friends should become unreal just because they present themselves in this way either. In the old times people communicated through letters, and it is conceivable that two persons who lived far away from each other could become friends entirely through correspondence. In his article criticizing Cocking and Matthews, Briggle cites Michel Foucault to show that letter writing offers a chance for both friends to reflect on what it is to be oneself: "The letter gives an opening for the other onto oneself, setting up a 'reciprocity of the gaze and the examination' (Foucault 1994, p. 216)" (Briggle 2008, p. 76). Communicating through letter writing is even of more limited dimensionality than emails, as it is possible to attach audio or video files with emails while this is not possible with old fashioned letters. Briggle cites Foucault who mentions Seneca and Lucilius whose correspondences were instrumental for the two to form close and genuine friendship, and this example can be supplemented by other examples of two personalities who became great friends only through correspondences throughout history, including in Thailand. Friends we correspond with through letters do not become unreal just because letters are the only form by which we communicate. So it should be likewise with online friends. In addition, in the case of Julius Caesar, there is certainly another sense of "real" where we say that the actor is a real actor. When we watch the play, we know full well that the Caesar we see on stage is not the real Dictator, but a performance of the play where the actor is representing his character. Furthermore, Briggle offers another argument where online friends can become more real than offline ones because the former can become more sincere whereas offline friends can be hampered by their physical closeness so that one might find it difficult to be sincere to the other (Briggle 2008).

Another main point in Cocking's and Matthews' argument is that, online interaction is a performance—one purposely build up one's outlook and profile in order to impress those who visit one's site. But there does not seem to be any compelling reason why this has to be so in all cases. It is more conceivable that friends who know each other well (either starting from offline life or not) usually open themselves up to each other. This is almost a necessary condition for genuine friendship.

This opening up does not go along with composing how one should look to the outside world. Furthermore, viewed from another angle, people always present themselves to the outside world. Even among close friends we do not appear naked to each other all the time and even friends have some secrets that one would like to keep believing that sharing them is rather irrelevant to the friendship. If this can be the case, then Cocking's and Matthews' argument that online friendship is unreal because people always consciously present themselves so that they look best online is not tenable, because, as we have just seen, people always present themselves to others as a matter of course in any world that they find themselves in (Goffman 1959), and among close friends there is more opening up of oneself to the other, as friends share more information to each other more than they share with the public, so there is less need for composing oneself.

In another article, Michael McFall (2012) also argues that online friendship cannot achieve the status of Aristotle's character friendship because character friends have to live together in close physical proximity in order to maintain their status as character friends. This is because living close together physically enables character friends to have single-filtered communication, the kind of communication that is not already interpreted by the sender and thus is more immediate, which is not available to online friends. Thus according to McFall online friendship can be only the other two types that Aristotle mentions, but not the highest and perfect form, genuinely virtuous, or character friendship. Online friendship is necessarily mediated by multifiltered communication, a kind of communication by which the sender has to interpret the message before sending it out to the receiver. Text communication is multifiltered in this way because the sender has to encode whatever meaning she would like to send in the form of language before sending it out. Suppose she would like to report an incident, for example, she cannot just send that incident over to the receiver or ask her friend to come over and witness it by herself, which would eliminate the need to report the incident in words. Thus the report has to be interpreted by the reporter. In sending out the message in text format, many things have to be omitted as it is not possible to include everything in the text. By asking the friend to come over to witness the scene of the incident herself, on the contrary, the friend thus can witness and soak up the scene through her five senses without the need for interpretation into text form. This single-filtered form of communication, according to McFall, makes for the more genuine and perfect form of friendship, one that enhances the moral and virtuous character of both friends.

Here is what McFall says on the issue:

> Though not a sufficient condition, living together is a necessary condition because of the level of perception and communication required. Character-friendship cannot be mediated entirely by technology because important perceptual and communicative elements of character-friendship cannot wholly be mediated technologically—especially concerning improving as moral agents through shared activity. (McFall 2012, pp. 223–224).

Living together is a necessary condition because without it McFall does not see that character friendship could develop. The "important perception and communicative elements" mentioned in the passage above are those enabled by single-filtered communication by denied by its multifiltered counterpart. An important feature of character friendship is that friends must improve their friends' moral virtue through their common activities, witness our discussion of shared activities and the similarity with the Buddhist view of spiritual friends in the previous section. Improving the friend's moral character requires, for McFall, that friends have access to items in each other's field of perception, those that are necessarily neglected or filtered out in the text-based communication that mostly links up online friendship. Even communication through the web-based video camera, as people do when they communicate on Skype, does not satisfy the requirement of single-filtered form of communication because even on Skype many are lost and do not get through to the receiver. A friend may report what has just happened to her. She might be there on the scene of the incident and uses her mobile phone to record what is happening around her. Nonetheless, for McFall this is still not sufficient because Skype video necessarily leaves out everything outside of its range, which might be necessary for the more complete kind of communication. In other words, Skype videos cannot compare with being there at the scene by oneself, and McFall insists that friends have be there physically to witness what is happening in person through all the five senses in order to have the necessary condition for their being qualified as real character friends. McFall says:

> Character-friends need single-filtered access to their other selves for pleasure, self-knowledge, and moral development. Multi-filtered communication does not simply blur perception. Rather, the difference between multi-filtered and single-filtered communication is that some images are radically distorted or not sent at all when mediated through technology, especially in the social and moral realm. Character-friends need access to the most accurate form of communication possible because they are involved in fine-grained moral development. Single-filtered communication is obtained when interacting physically with others, and character-friends must live together because they need considerable single-filtered communication. (McFall 2012, p. 229).

What is important is that text-based or videocamera-based forms of communication necessarily leaves out certain elements which are necessary for friends to maintain their virtuous relationship. The fine-grained moral development, according to McFall, would not be possible without living together in close proximity. However, if we pay attention to Aristotle's account of character friendship, we find that there does not seem to be any requirement that character friends need to be physically present together in order to maintain their status as character friends. This is the point we have already discussed in some detail in the previous sections of this chapter. Aristotle says: "Out of sight, out of mind" (literally: "A lack of converse spells the end of friendship"). What he is getting at here, as we have seen, is that friends need to maintain contact with each other, and the contact here does not seem to require the friends to live together, for there does not seem to be any compelling reason why genuine, character friendship cannot be maintained when the friends are far apart but keep constant contact with each other through the technology.

Here Cocking and Matthews, as well as McFall, agree that friends living far apart cannot maintain the quality of communication and common activities that are required for character friendship. That much is indeed the case, but it does not follow from this that friends who live far apart but keep on communicating with each other would automatically lose their status as character friends. The wish for well-being of one's counterpart can indeed be there even though the friends do not live together in close proximity. Cocking and Matthews, and McFall in his paper, agree that online communication is impoverished when compared with the offline type, and this impoverishment is the telling obstacle that destroys the friends' status as character friends. For McFall in the quote above, character friends need a lot of fine grained, single-filtered communication. But his argument would be cogent only if having fine grained, single-filtered communication is a necessary condition of wishing for well-being of one's counterpart, developing the other's moral virtues as well as those of oneself, as well as performing activities together. It is true that developing each other's moral virtues is facilitated greatly by the fine grained communication that McFall talks about. But in order to claim that this type of communication is necessary one needs to find a logical link between the two. Without single-filtered communication, no communication such as the friend's moral character will thereby be developed would be possible. However, we can certainly imagine a case where friends only communicate through letters or emails (or even smoke signals for that matter, though quite a lot of smoke would seem to be needed) and then the moral character of each is developed as a result. This is partly because, in the Aristotelian account, friends who can become character friends need to be virtuous already; otherwise each would not be able to recognize the virtuous character of the other so as to become friends with each other. When they are already virtuous to a certain degree, then, they would be able to focus on the aspects of the communication that are relevant and necessary for the development of one's own character and that of the friend.

Suppose I am somewhat virtuous and I find someone who is also quite virtuous, though not necessarily in exactly the same way (in the sense that Sherman explains in her paper that we have already seen). Then naturally that person and I would naturally become friends and it is likely that our friendship would develop along the line that Aristotle discusses in Book Eight of the *Nicomachean Ethics* that we have seen earlier. Since the intent of having and maintaining the friendship of myself and of my friend is identical, namely to develop the moral character and virtue of oneself and of the friend, then we would naturally focus only on the elements within the content being communicated that are directly relevant to the task. In this case, if my friend and I happen to be physically present in an environment together, since we are focused on developing each other's moral virtue, we would not be distracted by all the details surrounding ourselves that are there in the environment. For example, if we are walking together on a grass field watered by a small stream running nearby, we would focus our thoughts and attention of our conversation and its content, we would only occasionally notice the birds chirping in the tree and the cicadas that are making the noise. Like Socrates and his friend Phaedrus who are walking just outside of the wall of Athens conversing on the subject of love and friendship (See Plato's *Phaedrus*), we would stop to listen to the cicadas only when the insects happen to be a topic of conversation. In normal case friends who are intellectuals and who are interested in deep thinking would not stop to listen to the cicadas when the subject matter is something else altogether. Thus, McFall's requirement that friends need to have available shared physical environment and a way to perceive and talk about it that is rich and profuse would not be necessary. If Socrates and Phaedrus were talking about something else, say, the existence and objectivity of mathematical objects, it would hardly be possible that they would hear the cicadas chirping, even though they may be chirping loudly at the place where they are having the conversation. If friends are intently focused on some topic, it does not seem necessary, then, that they need single-filtered, fine grained communication. Only the kind of communication that transparently delivers what the friends are interested most, and when they are talking about mathematics or philosophy they are only interested in the abstract meaning, would be needed, and such kind does not need single-filtered communication because when the focus is already on the meaning, the stuff that is being communicated is already linguistic in the first place.

We might understand this point better if we compare the situation here with stage setting in a theatrical production. Ancient Greek plays are very minimal in their setting; usually the performance consists of three speaking actors wearing masks and a number of choruses. There are no elaborate sets that depict graphically the scene of the play. Everything is left with the imagination of the audience. A performance of *Oedipus Rex*, for example, did not need all the furniture and other trappings of a room in the palace of the King of Thebes; there is no need for golden chairs or the like. The stage is usually empty consisting of the bare minimum. And when there is a really graphic scene, such as when Oedipus gouges out his eyes after he finally learns the truth, the scene is not directly shown on stage, but related by a character who has just witnessed the event. The picture of Oedipus being overwhelmingly grief stricken, gouging out his eyes, is related by words alone through the mouth of the character, and the audience are expected to picture what is happening in their mind's eyes. This is sometimes more powerful than trying to show the scene directly, and it is, ironically, more realistic because practically speaking there is no way to depict the scene of one gouging out his eyes on stage without some make belief or pretense. Whereas when the audience listen to the character telling them what he has just seen—Oedipus crying out with deepest grief and putting a dagger into his eyes, they can readily imagine that Oedipus is *actually* doing this in the adjacent room. Sometimes the power of words is stronger than whatever an elaborate prop and setting can achieve. Thus it is also the case that one can grasp the meaning being conveyed by the two friends better if the meaning is conveyed through text alone. Without the stimulation provided by graphic forms, understanding the words' meaning can excite the imagination much more powerfully. *Oedipus Rex* is generally known as a moral play, its message being that no matter how much one strives to be good, there is always an element of luck that no man can escape. The deep irony in the play is that Oedipus is a very moral person, a very good man, but the gods play tricks on him so that no matter how hard he tries to be moral, he unwittingly finds himself having committed the most heinous deeds imaginable, killing his own father and marrying his own mother. All these deeply moral points are sufficiently conveyed through the power of Sophocles' masterful words alone, on a very bare Greek stage consisting of nothing but actors standing around acting out their parts.

Hence, if this is the case, then it seems that textual communication (such as what is happening in the production of *Oedipus*) can achieve the goal of moral deliberation and reflection of the highest order. This should apply to genuine character friends who find out that they have to communicate only by emails too.

I have tried to argue, then, that text-based form of communication can be a medium through which the highest form of friendship in Aristotle's sense can develop. In another paper, Munn (2012) argues that such friendship can develop in either the physical world or in the immersive virtual world which look virtually indistinguishable from the physical world. He argues further that through the medium of communication alone character friendship cannot develop because shared activity cannot take place in that medium. According to Munn, shared activity is one where "friends engaged in such activity jointly pursue a goal when all of them not only desire a particular outcome, but also desire that the outcome be the product of the combined activity of the group, as it is composed" (Munn 2012 p. 4). Friends engage in shared activities, according to Munn, when they form a group and when they specify a goal that is shared by all the members of the group. It is important that the group shares and works at achieving the goal together. If someone proposes that a member is dropped and someone else from outside of the group brought in order to increase the chance of achieving the goal, then that would defeat the purpose of having the group of friends in the first place. The shared activity and the common goal that the whole group aims at achieving as this particular group is essential. Munn then argues that this kind of shared activity is not possible in the older form of predominantly text-based form of online social networking and communication. Only in the physical world or the immersive virtual world which eliminates most of the boundaries between the two suffice to promote this kind of friendship.

Munn says that the shared activity must be a kind where friends "desire that the outcome be the product of the combined activity of the group" (Munn 2012, p. 4). Furthermore, friends who join in the shared activity must enjoy the activity when shared with her particular friend here and no one else. One who enjoys riding the bicycle with a company no matter who she is does not qualify for the shared activity here. For two friends to engage in the shared activity in this sense they must recognize the uniqueness and identity of the other, and these are necessary for the particular shared activity (Munn 2012, p. 3). Here communication and activity are different according to Munn. Communication is needed for the stage of planning the activity and talking about it among friends, but they cannot substitute for the activity in itself. Friends who love riding the bicycle together may engage in a lot of communication about when their next bicycle trip might be, how they enjoy a ride so much and so on, but that is different from the actual act of riding the bike. Since the latter cannot take place in the text-based online forum, Munn then concludes that either the physical world or the immersive virtual world is sufficient for this kind of shared activity to develop among friends.

However, when we consider the case of friends who live far apart and keep on corresponding with each other through the old-fashioned letter, the kind of shared activity that both are engaged in is naturally limited. It consists only in writing the letter, putting it in an envelope, sending it out, receiving a reply letter, opening it, reading it, keeping it, reading it again, composing a reply, and so on. These are the activities that both friends have to engage in. One might say that these activities do not take place together, but if the friends are really serious about letter writing, the activities surrounding it can be quite time consuming. A friend might talk about his activities in the letter that he sends to his counterpart. He may be fond of gardening, spending quite a significant amount of time outdoor tending to his plant, and then spending more time writing about it by hand, sharing it with his friend, who might also happen to be a gardener too. Hence at least the two friends here share some activities together, only that they do not do their activities in close proximity with each other. Compare the activities that these two letter writing and gardening friends do with the example of two companion bicycle riders in Munn's article, I don't see any substantive difference that would bar the former friends from achieving the status of character friends in Aristotle's sense. The bike riding friends do not ride their bikes with anybody else, and so too are the two letter writing friends; they just do not write their intimate letters to anybody else, thrusting their letters into the public domain for everyone to see. We have seen, moreover, that text-based communication, what McFall calls multi-filtered communication, can convey moral deliberation and reflection of the highest order; thus there does not appear to be any reason why these letter writing friends should not qualify as real character friends.

In another paper, Johnny Søraker (2012) argues that virtual friendship (the kind of friendship that exists in virtual world and not merely on social networking sites) can afford what he calls "prudential" value, that is, the value that is useful or pertinent for some people at not necessarily everybody. Those who have little or no access to physical friendship can have their lives fulfilled at least to a certain extent through virtual friendship enabled by information and communication technologies, so at least some value engendered by virtual friendship is still better than no value at all for those who happen to lack the access to physical friends. Søraker cites a number of empirical studies that examine the values and effects of friendship and found that friendship is essential to psychological well-being (Søraker 2012, p. 210). This shows that at least virtual friendship is of some value; in other words, it is still better than having no friends at all. Instead of replacing the traditional forms of friendship, virtual friendship, according to Søraker, tends to augment or exist alongside the already existing form of friendship rather than replacing it totally. Søraker again cites a number of empirical studies in support of this point (Søraker 2012, pp. 201–211). Søraker does not discuss text-based form of friendship much in his paper. He agrees with Cocking and Matthews that the limitation imposed by this form of communication, the lack or reduction of involuntary self-disclosure, may result in virtual friendship being less able to deliver the prudential value of friendship to certain people because virtual friendship (and by extension online friendship of the kind we are discussing) is of more limited dimensionality than physical friendship. Søraker says that online friendship can be mediated only through sight and sound, as these two senses are well suited to digitalization found in today's computer. The other three senses—taste, smell and touch—are much harder to digitize and transmitted over the network, so virtual or online friendship is constrained by the limit of the current technology. As I have said, being limited in means of perception does not have to result in the friendship being of lesser quality, as imagination can fill in the gap which could be even more powerful than having fuller access to the senses. All things being equal, having access to two senses is of lesser quality than having access to five, but if friends who are mediated by two senses make *full* use of these two senses, especially when the senses are used to convey intense meanings, then the quality of the friendship can be much enhanced.

5.4 Online Self and Online Friend

In an article published in *Ethics*, Dean Cocking and Jeanette Kennett (1998) argue that there are two views that provide a philosophical account of friendship, viz., the secret view and the mirror view (quoted in Ray Pahl 2000, p. 80). According to the secret view, friends share secrets with each other. It is the trust that each friend has that cements the bond between the two. The more secrets are shared between the friends, the closer the tie is between them. On the other hand, the mirror view holds that friends are like mirror images of each other. This is, on the surface at least, what Aristotle's view of friendship looks like as we have seen. Virtuous, character friendship is that between friends who are already virtuous. However, development of virtue and mutual support would not be possible if friends were entirely alike; thus there has to be a dynamic between likeness and difference among the friends—not too much of each but not too little either—in order that friends maintain the dynamism that could result in the realization of the mutual support. This is in fact the third view of friendship, which Cocking and Kennett dub the "drawing" view. According to this view, friends are sufficiently different from each other such that one can be receptive to the characteristics that define the other and transform one's characters and interests to suit those of the friend too, and the same thing happens on the friend's side. In this way, friends can be see things from the perspective of the friend herself. For example, if my friend loves opera and invites me who has never been interested in the genre to attend a performance, then I might accept the invitation because in some way her interests have an effect on me so that I begin to look at the world more through her eyes. In this case I am not necessarily becoming more like my friend, which would be in the "mirror" view, but the main point is rather than I am responsive to her interests and there is a kind of dynamism going on between the two of us. According to Cocking and Kennett, having "one's interests and attitudes directed, interpreted, and so drawn … is … both typical and distinctive of companion friendships, yet has been largely neglected in philosophical literature on the subject" (quoted in Lippitt 2007).

According to Aristotle, my friend is my "second self" or my "other self." This view goes well with the mirror view described above. Friends are like each other; that is way they become friends in the first place. The highest kind of friendship, one where the reason for the two persons to become friends is neither pleasure nor utility, but virtue alone, is for Aristotle one where the two friends are equally virtuous.

Being my "second self," then, seems to imply that my friend is just my self extended or copied onto the body of the friend, and the same of course goes with me who is the copy of my friend. In fact the three views on friendship that we have seen here are related very much on the interplay between the friend and the self. For the mirror view, the friend is one's second self. For the secret view, it is not quite as obvious, but one can presume with good reasons that friends with whom one readily share all the secrets and all the deepest thoughts and beliefs must be sufficiently like each other that the view closely follows the mirror view in terms of the relation between friends and selves. After all, if, ideally, my friend and I share all the deepest thoughts so that there is no privacy among both of us, then as the self is constituted through information as we have seen in Chap. 3, then this ideal sharing would result in the two of us becoming just perfect copies of each other. For the drawing view, however, the picture is a little more complicated. Since my friend and I are not exactly the same as each other, the differences between us have to result in some differences among our two selves. Nonetheless, we cannot be completely different; otherwise there would not be anything that binds us together as genuinely virtuous friends. It is, then, the dynamic interplay that exists between us that is at issue. Furthermore, for the drawing view what constitutes my own self can be influenced and transformed by the relation I have with my friend. The decision I make to attend the opera in order just to follow my friend and to be with her, to enjoy her company, and the opera too, reflects a change in my personality as a consequence of my drawing of the character of my friend as well as drawing of my own makeup as a changing self too.

The three views offered by Cocking and Kennett are those that attempt to characterize the kind of genuine character friendship that Aristotle talks about. However, when it comes to online friendship the picture remains the same. What I would like to get at here is that the three views here can also explain, each in its own way, how the situation of genuine friendship in the online world should be understood. Online friends can and do share inner thoughts and secrets to each other; in fact so much so that privacy becomes a concern as is well known. They could be viewed as completely mirroring each other, which they will actually become if the sharing of inner thoughts between the two is complete. They also, according to the drawing view, be seen and understood as sharing thoughts among themselves, but in such a way that each personality remains individualized and dynamically unique; that is, not statically unique in the sense of not being able to change through influx of new information at all, and not changing so much and rapidly that no uniqueness and individuality can be discerned.

It is the drawing view and its effect on how to understand the self that is more challenging. How can friendship have an effect on a conception of the self? And when the situation happens in the online world, does that have an effect on either the characteristic of friendship or of the self or on how friendship affects the self? As we have seen in Chap. 3, the self, ultimately speaking, is only a construction out of a large number of mental and bodily episodes that work as a fulcrum point for having a stable point of view and a sense of individuality inside the world (See, for example, Metzinger 2009). So when the self interacts with a friend, then the myriad of episodes are then shared among the two. The secret view gives us the clearer picture of how the sharing is constitutive of the friendship than the other two views. As I have just said, when two friends share their innermost thoughts and bare themselves to each other, privacy falls by the wayside. It is not that the two friends do not respect privacy any longer, but they trust each other enough to share these thoughts with each other. We find then a fusion of the two selves. If the self is constituted through information as previously argued, then sharing all these secrets would be tantamount to merging the two selves together, so in fact the secret view becomes fused with the mirror view. That is, one view ultimately logically implies the other and vice versa. According to the Extended Mind Thesis, the mind can extend beyond the confine of the physical body; thus there is nothing in principle that would prevent the minds of two friends from extending toward each other thereby fusing the two minds together and perhaps working as one. This may sound far-fetched, but we can imagine a situation where two friends are so in tune with each other that they are metaphorically speaking "of one mind." This situation does not require any sophisticated or futuristic technology. It only requires two friends whose minds are working together closely and the two friends perform as one team going as if they were one unit while in fact they are not physically speaking. An example that comes to mind is a team of double players in tennis or in badminton. They have to work together very efficiently as one unit in these fast paced games. There are always certain moments, however, when the fusion and the communication break down and very often the team loses a point as a result, but they always aim at creating a team, one seamless unit whose components work together for the common goal. In this situation, sometimes one player "knows the mind" of the other such that they anticipate what the other would do and react to that even before the act itself happens. This happens often in double competition. In this situation where the seamless team is created, we could say, metaphorically, that the two friends or the two double partners fuse their minds together. Furthermore, there is nothing in principle that says that this kind of intense collaboration, this fusion among two selves, cannot happen online. Certainly it is not possible for the two friends to play double tennis online. What they can do, on the other hand, is to play computer games together as a team and here they can merge together as intensely and as closely knit as the team that play tennis on the court.

Friends are in fact very important in the social network. They are the *social* in the social network. We have seen that according to the Extended Mind Thesis, friends can fuse their minds together at least temporarily when they are engaged in some intense collaborative work together such as playing double in tennis. Moreover, it is our friends that define who we really are in the social network. Imagine one who is completely alone on Facebook. In that case Facebook would not be the social network that we are familiar with. There will be no conversations, no sharing with anyone, no giving and taking of barbs and compliments and so on. In fact we can imagine a scenario where one enters Facebook with multiple email addresses but all these addresses belong to this one and the same person. Here it is possible that there are more than one person in his account and timeline. The person is in fact befriending himself and is engaging in the activities on the social network site with himself.

He may disguise himself to be this or that particular person, but we would feel that there is something seriously wrong with his mental makeup in doing this. People need real friends on the social network. We have just discussed the situation where friends work together so closely that they become one unit, so to speak. On the other hand, we also need friends to maintain our own identity and uniqueness among the nodes in the online social network. As in the case of offline friends that we have seen, we need friends in the online world to maintain our identities; otherwise we would become like the one who enters the social network alone and makes friends with himself. In that case there is no way for him to learn who he actually is because there is no feedback from someone else so that he can look at himself through the eyes of another. This is what the drawing view of friendship emphasizes as we have seen. In addition to friends transforming themselves through their interaction and communication with their friends, each friend, each person in the friendship network, finds out who he is in the network through the reaction that his friends give of himself. This is in fact an illustration of the view that we can only know who we are through our membership in a community. In the world where one lives alone, as in the situation where one does not befriend anybody else on Facebook, there is no check as to who one can be. In that world the person can make up himself to be Napoleon Bonaparte and his "friends" (who are in face himself registered through other means) are those who reinforce this conception. Thus his "friends" might be the Empress Josephine, his brothers, his Prime Minister, and so on. Even if "Josephine" or his "brothers" are in fact others who play along with "Napoleon," this does not improve the situation as the playing along has to create a make believe world bearing no relation to the real world outside. The important aspect of friends who reflect who we are is that we cannot control them. They are aspects of the external community on which we have to depend for our livelihood and our being. This can happen in the online as well as the offline world. As in the case of the drawing view, furthermore, it is not that each one of us has a fixed identity that does not change, but our personalities, our sense of who we are, can and do change in accordance with changing circumstances. Here our friends do change who we are, but as we are friends to them we do change who they are also.

5.5   Conclusion: A Critical Look—What Does the Extended Self View Offer?

We have seen that Cocking and Matthews argue that online friends are always inferior to offline ones because of the limitation of the technology that mediate interaction among online friends. However, when it comes to moral activity, which for Aristotle is what makes for the highest form of friendship, there does not seem to be any reason in principle that such activity cannot take place online. The setting of ancient Greek tragedy is minimal in the extreme, yet the form is widely regarded as being able to convey and illustrate deep moral problems, as Aristotle himself argues in detail in his *Poetics*. The idea is that when conversation is reduced to text only, one makes fuller use of one's imaginative power, thereby strongly engaging the cognitive power. Thus, the deficiency that is there in the environment of online friends can turn out to be a positive force, since text-only communication forces the friends to make full use of their imaginative and cognitive power. Moreover, the highest form of human ideal, such as the idea of universal love as we have seen, is also possible through online communication. If such highly ideal form is thus possible, then there is nothing that distinguishes the offline and online forms from each other. This view is also supported by Nietzsche's analysis of friendship as we have also seen.

However, this does not imply that online friendship is equal to offline one in other aspects. Writing in 2007, when Facebook is just starting up at Harvard campus, and when the dominant social network sites are MySpace and Friendster, Christine Rosen writes:

> The structure of social networking sites also encourages the bureaucratization of friendship. Each site has its own terminology, but among the words that users employ most often is "managing." The Pew survey mentioned earlier found that "teens say social networking sites help them manage their friendships." There is something Orwellian about the management-speak on social networking sites: "Change My Top Friends," "View All of My Friends" and, for those times when our inner Stalins sense the need for a virtual purge, "Edit Friends." With a few mouse clicks one can elevate or downgrade (or entirely eliminate) a relationship (Rosen 2007, p. 27).

One has a lot of choices and power in "managing" one's friends on a social networking site than in real life, where in many cases one does not even have much choice in choosing who is going to be one's own friends. In offline life, one finds that one is, in Heidegger's word, "thrown" *(geworfen)* into a particular place and time. I find myself a Thai born in Thailand, and so on; thus my choice of friends is limited to this situation. However, Rosen says that the user can edit their friends, adding them and deleting them as they see fit. This kind of "bureaucratization" of friendship is unavailable or impossible in offline life. The picture conveyed is then that of one being able to get rid of Heidegger's "thrownness" situation all together. One is totally free of one's relation with one's cultural universe; one can construct one's identity with no constraints at all. If friends are parts of one's cultural universe and parts of one's sense of identity, then the idea that one can construct one's identity with no constraints means that one can add, delete, thereby "managing" one's friends also with no constraints. It is as if everything in the universe revolves around oneself. This is reinforced by the experience of the user, who typically sees their online friends only as pictures on the monitor whereas they themselves are real life, breathing user with physical body. What is missing is certainly that the friends are also real life users who see their own friends as pixels on screen too. However, if we see that the online and offline worlds are not strictly separate, but are continuous with each other, then one begins to see oneself as an online friend to the friends that one has too. This tendency to view one's friends as only pixels perhaps feed into the perception that one is the only thing in the universe around which everything else revolves. But if one sees oneself as another online friend to the group of the friends in the network then one might start to see oneself more as a node in the network rather than the absolute center. Once that happens, the bureaucratization that Rosen talks about here perhaps loses much of its force, because the management that one is making would be not too much different from the management that people also do in offline life, when they write up their own list of friends in their address book, adding and deleting those names to suit the changing circumstances. For example, those names in the list which we have lost contact with for a long time are likely to be deleted, or if they are still kept in the list they may be ignored most of the time. And when we look at those names again we may be reminded of the times we spent together with the owners of those names. We may be tempted to give them a call to find out what they are doing. In the online world we of course also have a means to reconnect with long lost friends (if they still have a presence online or within our social network, of course). In either case, the management of our friends list in the online and offline worlds (such as in a notebook) are quite similar.

References

Aristotle. (1962). *Nichomachean ethics* (M. Ostwald, Trans.). Indianapolis: Bobbs Merrill.
Briggle, A. (2008). Real friends: How the internet can foster friendship. *Ethics and Information Technology, 10,* 71–79.
Cocking, D., & Kennett, J. (1998). Friendship and the self. *Ethics, 108,* 502–527.
de Montaigne, M. (2003). Of the art of discussion. In *The complete works* (D. Frame, Trans.) New York: Knopf.
Cocking, D., & Matthews, S. (2000). Unreal friends. *Ethics and Information Technology, 2,* 223–231.
Foucault, M. (1994). In P. Rabinow (Ed.), *Ethics: Essential works of Foucault 1954–1984* (Vol. 1). London: Penguin.
Goffman, E. (1959). *The presentation of self in everyday life.* New York: Anchor.
Hongladarom, S. (2011). The overman and the arahant: Models of human perfection in Nietzsche and Buddhism. *Asian Philosophy, 21*(1), 53–69.
Kierkegaard, S. (1991). You shall love your neighbor. In M. Pakaluk (Ed.), *Other selves: Philosophers on friendship* (pp. 233–247). Indianapolis: Hackett.
Lippitt, J. (2007). Cracking the mirror: On Kierkegaard's concerns about friendship. *International Journal for Philosophy of Religion, 61,* 131–150.
McFall, M. (2012). Real character-friends: Aristotelian friendship, living together, and technology. *Ethics and Information Technology, 14,* 221–230.
Metzinger, T. (2009). *The Ego Tunnel: The science of the mind and the myth of the self.* New York: Basic Books.
Miner, R. C. (2010). Nietzsche on friendship. *Journal of Nietzsche Studies, 40,* 47–69.
Munn, N. J. (2012). The reality of friendship within immersive virtual worlds. *Ethics and Information Technology, 14,* 1–10.
Nietzsche, F. (1990). *Unmodern observations* (W. Arrowsmith, Trans.). New Haven: Yale University Press.
Pahl, R. (2000). *On friendship.* Cambridge: Polity.
Panaïoti, A. (2013). *Nietzsche and Buddhist philosophy.* Cambridge: Cambridge University Press.
Rosen, C. (2007). Virtual friendship and the new narcissism. *The New Atlantis, 17,* 15–31.
Sherman, N. (1987). Aristotle on friendship and the shared life. *Philosophy and Phenomenological Research, 47*(4), 589–613.

Søraker, J. (2012). How shall I compare thee? Comparing the prudential value of actual virtual friendship. *Ethics and Information Technology, 14*(3), 209–219.

Stone, B., & Frier, S. (2014). Facebook turns 10: The Mark Zuckerberg interview. Bloomberg BusinessweekTechnology. Retrieved from http://www.businessweek.com/articles/2014-01-30/facebook-turns-10-the-mark-zuckerberg-interview#p3

Chapter 6
Computer Games, Philosophy and the Online Self

In addition to the social network, the online self has found itself in the various genres of computer games too. A main characteristic of a computer game is that the player can control an aspect of the game, and when games become more sophisticated, these controllable elements becomes more enriched, thereby becoming another form of the online or virtual self. From its starting point as a simple game like *Pong*, where the player moves a virtual paddle and tries to hit a ball back to the opponent, the players today can become immersed in a virtual reality where they control their "avatars" in a detailed environment that is graphically rich and sophisticated. Whereas *Pong* very roughly simulates the reality of a ping-pong table, games like *SimLife* aims at presenting a total simulation of a real life of a person from the moment when she is born until she dies, and during this life the player encounters several events that closely imitate what one would expect to find in one's real life. And when SimLife becomes an online platform, allowing players from various locales to engage with one another through their playable avatars, the game itself becomes a social network, albeit a virtual one that is different from Facebook in that it does not aim at representing what is actually there in reality. I can create a character in *SimLife* which does not resemble what I look like in reality at all and then *become* this character in the online social platform. Online games such as *Club Penguin,* a game designed for children, or *League of Legends,* a very popular online fighting game that is being played by millions throughout the world, thus present an interesting phenomenon that merits close examination and analysis. What is the relation between the real life player and her avatar? What is the difference, if there is any, between rich avatars such as those in *SimLife* and the very simple playable character one finds in arcade games such as *PacMan*? Is it plausible to say that when one is playing a game through an avatar one is in fact being "immersed" in the environment of the game? What accounts for the identity of an avatar or a playable character in a game? We will try to unpack these questions in the course of this chapter, relying of course on what we have learned from the previous ones especially the analysis on the identity and constitution of the online self.

© Springer International Publishing Switzerland 2016   147
S. Hongladarom, *The Online Self*, Philosophy of Engineering and Technology 25,
DOI 10.1007/978-3-319-39075-8_6

The idea to be argued for in this chapter is that the Extended Self Thesis can explain the relation between the player and her avatar in a game environment in such a way that avoids most of the problems that have plagued other conceptions. The avatar in a game is not merely a puppet that is totally controlled by the player. The avatar does, in a way, have her own thought and consciousness (especially if we conceive of avatars that are remotely controlled, such as the ones in the movie Avatar, by the players themselves). This can be explained most plausibly, as I shall argue, by the Extended Self Thesis, On the other hand, the avatar is still distinct from the person of the player; it is not the case that the player loses her own identity through her playing or assuming the role of the avatar. We shall see how the argument for this claim unfolds in the course of the chapter. Furthermore, there is a new possibility afforded by the new brain-to-brain integration technology. When two brains or more are connected in the same network, new avenues will be open in the context of games and other areas too. Perhaps there will be completely new kinds of games, or the lines between games and real life will be even fuzzier that they are now.

6.1   What Is an Avatar?

The word 'avatar' comes from Sanskrit *avatār* meaning 'coming down.' It refers to an action of a god (in almost all cases it's Vishnu) who assumes the form of a mundane being in order to fight with evil characters to restore peace to the world. In Hindu mythology the three supreme gods have different functions. Brahma is the creator; Vishnu is the preserver and Shiva is the destroyer. Thus when Brahma creates a world, it is the duty of Vishnu to preserve it, bringing peace to it and restore the cosmic order, and when the time slot of that world is up, Shiva will open his third eye which will destroy everything, opening up space for Brahma to create another world again, thus continuing the cosmic cycle. What is of interest to us here is the act of Vishnu in coming down and assuming the form of a mortal creature. In the mythology Vishnu has come down to this world ten times already. He first took an avatar as a fish, then as a tortoise, a boar and a half man/half lion called Narasimha. Then he became a dwarf, a warrior with an ax; then he becomes incarnated three times as a human being (as Rama, Balarama and Krishna), and then in his final avatar he appears as a man on a white horse, Kalki, the harbinger of final judgment and heralder of the end of the current age of the world.

When Vishnu decides to take up an avatar, he simply enters the womb of the woman who will be the mother of the avatar. In this way he, as a god, takes on human form. The canonical literature is rather silent on the issue whether there is any kind of separation between Vishnu himself as a supreme god and his avatar as either an animal such as a boar or a human being. That is, during the time of an avatar such as when Vishnu comes down to earth as Rama, who is a human prince, are there two beings, Rama and Vishnu? Or is it rather that Vishnu himself disappears while Rama is around? However, since Vishnu is a supreme god, presumably he has a lot of power to do many things; hence remaining as a god in heaven while at the same time assuming human form on earth would not be much of a trouble for him. In any case, although that remains a possibility, in many passages in the literature Rama himself (or other avatars of Vishnu) retains his characteristic as a god. That is, while he assumes the human form, he is sometimes described as a god and he has all the supernatural powers that befit a god. In other words, when Vishnu comes down to earth he does not come down as *only* a human being (or an animal), he is still a god who appears among earthly creatures both as a god and as one of them. In this case, then, Vishnu's coming down can also be described as an incarnation, in the same sense as Jesus is an incarnation. In Rama (or in other avatars or incarnations), there are both the nature of the god and of the earthly creature united in one being; likewise, Jesus is also a union of God (he himself being the second God of the Trinity in his own right) and of a human being. Jesus is "the Word made flesh," according to the Gospel of John (John 1:14). In this sense, then, Jesus can also be considered an avatar of God.

It is this aspect of the avatar that is relevant to game playing. Rune Klevjer, in his Ph.D. dissertation on the topic, writes:

> An avatar is an instrument or mechanism that defines for the participant a fictional body and mediates fictional agency; it is an embodied incarnation of the acting subject. It is dependent on the principle of the model, and acts as a dynamically reflexive prop in relation to its environment. Its capabilities and restrictions are based on the objective properties of the model, and these capabilities and restrictions define the possibility-space of the player's fictional agency within the game. The avatar therefore defines the boundaries of embodied make-believe (Klevjer 2006, p. 87).

What is missing in Klevjer's description is perhaps the connection with a god. However, we can also see that for a participant to take part in a game of make-believe such as the one appearing on a computer monitor, the participant, metaphorically speaking perhaps, is taking on the role of a god such as Vishnu who, out of the desire to restore cosmic order (which could be an objective of the game), takes on the body of an "earthly" creature (what is happening inside the game being seen from the eyes of the god as "earthly"). In this way the participant, or the player, "mediates fictional agency," or in other words manipulates the avatar. However, a difference between a god taking on an avatar and a human player doing to same is that a god never loses. Vishnu always defeats his opponents and thus the universe is preserved. The same cannot be said for the human participant. What is more relevant, nonetheless, is that when Vishnu takes up a form of an earthly creature in an avatar, he enters a different environment from the one he lives in as Lord Vishnu. This is the essential aspect of an avatar. His second avatar, that of a tortoise, happens when the entire world system is going to fall apart because of a lack of foundation. Thus he takes the form of a giant tortoise and put the whole world on top of his shell so that the world does not disintegrate. The environment of the ocean when the tortoise lives and the world system that rests on the tortoise's shell is a far cry from the heaven where Vishnu resides with his consort Lakshmi. An avatar always finds himself or herself in a totally different environment than that of the original person. Many game theorists regard this *other* aspect of the avatar world as fictional, though, as we shall see later on in the chapter, there can be some game environments where the notion of fiction does not always apply.

6.2   Relation Between the Player and the Avatar

One of the most important features of computer games is that the player can control an element in the game to engage in the play and achieve the objective of the game itself. A very simple game such as *Pong* has the player control a simulated paddle with the aim of not allowing the ball to slip off the screen. Here the relation is between the player and the paddle. Another, more recent game, *Marble Madness,* has as its aim the control of a marble on a variety of obstruction courses, and the objective is to keep the marble on the board as long as possible and arrive at the destination point. The player controls the marble either by the mouse or the keyboard. In these games, the player controls an element, either the paddle or the marble, and tries to achieve the objective of the game. In other genres, such as role playing or adventure or first-person shoot out games, the player becomes merged with the character inside the game. The player still controls the character, but there is the added sense of the player herself becoming identified with the character who she is playing as. A result of this apparent merging is that instead of the player staying outside the game controlling what is happening in the game, the player/character controls what is happening inside the game *as if* she is already inside. Here the main philosophical problem is whether it makes sense at all to talk about oneself being inside the game instead of just staying outside and playing the character as if the latter is only a puppet. According to the Extended Self Thesis, I would like to argue that the self of the player does indeed extend toward the context of the game. Thus the playable character or the avatar in the game is not a puppet, but the character is not the whole self of the player either. In "Enter the Avatar: The Phenomenology of Prosthetic Telepresence in Computer Games" (Klevjer 2012, pp. 17–38), Rune Klevjer, however, argues for the concept of prosthetic telepresence where it makes sense to talk about the player being actually present inside the game. Taking a cue from Merleau-Ponty (1962), Klevjer argues that the puppet that functions as an avatar in the game does represent the body of the player now being transported to the context of the story of the game world. The "body-subject"–the subjective feeling that one has, the sense of being here and there, is manifested, when playing the game, inside the game as someone who is a character there, and the "body-object"– that through which the body-subject is manifested—is then the objective embodiment of the avatar inside the game. Thus for Klevjer both the subject and the object, which for Merleau-Ponty are both based on the body, are transported through what he calls "prosthetic telepresence" to the context of the game.

In order to illustrate this point let us look at an example. It is quite easy to imagine a game where the player becomes Don Quixote, whose aim is to fight a windmill. Here the Don Quixote that one finds on the computer screen in the context of the game is the body-object and when someone plays the game intently, one loses herself and focuses exclusively on the task of fighting the windmill, the subject sense of someone who is Don Quixote *is* then the body-subject. For Klevjer this shows that the player finds himself or herself inside the game completely, through the agency of prosthetic telepresence. Then Klevjer has the following to say:

> This means that our experience of being taken into the game world by our avatars can be explained without recourse to fictionality. Undoubtedly, make-believe plays an important role, insofar as computer game marionettes would also be conceived as humanoid agents or characters who somehow acts on our behalf. Nevertheless, proxy embodiment is a trick at the level of the phenomenology of the body, not a trick of fiction. The sense of bodily immersion that is involved in avatar-based play is rooted in the way in which the body is able to intuitively re-direct into screen space a perception of itself as object, which is the perception of itself as part of external space. A mouse cursor cannot function as a proxy in this way, not because it lacks fictional elaboration, but because it has to objective present within screen space (Klevjer 2012, p. 29).

Klevjer does not agree with the usual view that telepresence in a game is a function of the game's fictionality. Being Don Quixote in the game is not a result of the story in the game where Don Quixote is a main character. Thus for Klevjer playing Don Quixote in the game is not the same as playing the role of Don Quixote in a play. The difference is that in the play there is no "material divide" which for Klevjer is a divide between the actual world outside of the game and the world inside the game. The actor playing the role of Don Quixote on stage does not experience this divide because the world according to the story that is being played out on the stage is on the same side of the material divide as the real world. The actor, when he plays the role of the Don, is not engaged in telepresence of any kind, nor is he attached with any prosthetic device that transports him to another world. The world in the play in a make-believe world; the audience are asked to experience what they see on stage as the real Don Quixote taking on the real windmill (even though the windmill on the stage may be made of paper). However, in the game context, the player becomes Don Quixote and is transported to the game world through the avatar which the player controls. There is a divide here between the player herself and her avatar, whereas in the play the actor only plays the role of the Don, but he is not directing any avatar who is located outside of his own body. Klevjer relies on Merleau-Ponty's analysis of the phenomenology of body perception to show that the avatar is a telepresence of the player—the player is actually inside the game. This is why Klevjer talks about "a trick at the level of the phenomenology of the body, not a trick of fiction" (Klevjer 2012, p. 29).

In another related paper (Linderoth 2005), Jonas Linderoth argues that there are three roles for the avatar in the game, namely avatars as roles, tools and props. Avatars as roles mean that the avatar become a fictional character whose identity the player assumes during the course of the game. Thus playing Don Quixote would exemplify avatars as roles here. Avatars as tools mean that the avatar is used as a means by which the player extends herself into the context or the story of the game. This is perhaps the most common use of the avatar, and is akin to Klevjer's analysis of the avatar as a puppet or a marionette. The avatar as tool is that by which the player plays the game and achieves the objectives set by the rules of the game. The third use of the avatar, avatars as settings, means that the avatar is taken to be a means by which the player presents his or her sense of self to the outside world. Thus this is the most interesting use among the three here. Linderoth mentions a case where a boy takes on an avatar of a female warrior, knowing full well that she is of different sex from him. When asked by his friend, he said that he would like to try the avatar anyway because she is "awesome" (Linderoth 2005). Although he chooses the female character this does not translate into a change in his own personality; on the contrary he uses the female warrior avatar for her prowess in fighting and thus, according to Linderoth, only the fighting ability of the avatar is paid attention to. Linderoth says that in this case the avatar ceases to be her own character and becomes a "prop" in the presentation of the self of the player himself, who would like to project his fighting ability on the game, bending the outward form of the avatar in the process (Linderoth 2005).

In his paper, Linderoth does not agree with the idea that the player becomes immersed in the world of the game during the play, which is contrary to Klevjer's argument that we have seen. According to Linderoth, it is a "holodeck myth" (Linderoth 2005). Instead of having oneself immersed in the context of the game, Linderoth, following Fine (1983), divides the layers of the player-avatar relationship into three levels—the level of commensensical reality, the level defined by the context of the game, and the level of the larger social context in which the player finds herself in. These three levels intermesh with one another and for Linderoth it is not just a matter of a player finding herself immersed in the game only. Thus in my game of Don Quixote fighting the Windmill, the level of common sense reality would be the level where one finds oneself in real or offline life—one plays a computer game at home, for example. At Level Two, there is the world of Don Quixote de la Mancha, his sidekick, the Windmill and so on—in short the world inside the story of the game itself. Level Three is where the avatar becomes part of a setting that allows the player to present a sense of himself or herself. Thus in choosing the female warrior character, Linderoth's 8-year-old game player thus has a means to present himself, what he would like to world to perceive, in such a way that is markedly different from what is assumed in the female avatar. To do this it is necessary that external socio-cultural contexts be considered, which is the aim of Level Three from the beginning.

So instead of enabling prosthetic telepresence, Linderoth's avatar allows the player to assert his or her own identity in the offline world, the one that has been with him or her before coming to the game itself. Playing the role of an avatar in the context or the story of the game is only one part of what a player can do according to him. However, we have seen that there is a difference between traditional role playing as acting a part in a play and playing an avatar in role playing games where one seems to become identified with the avatar. In the former, one becomes the character being played (such as Hamlet or Don Quixote), and in the latter one controls an avatar while maintaining one's own individually subjective presence as oneself outside of the game environment. Linderoth apparently thinks that there is no case of one being immersed in the game environment and actually becoming the avatar, and the most the player can become is to act out or to pretend to be the avatar as one does a character in a play. Klevjer, on the contrary, maintains that prosthetic telepresence is the key to understanding the relationship between the player and her avatar, and "telepresence" of course means that one is transported to the game environment, and he also states that fiction has to role in the relationship here as we have seen. I believe that Linderoth's account is more tenable here. It seems that as long as one is aware of one's bodily stance as someone who is playing the game in the offline world, a bodily presence that needs nourishment and oxygen to survive as a physical organism, then the bodily presence cannot be dispensed with lightly. Being immersed in the game environment through an avatar sounds like one can forget about one's bodily condition as one who sits in front of a computer interacting with the game through the controlling device and somehow is transported to the game world. Even when Klevjer talks about "camera body," i.e., the unseen body whose perspective is there in the game so as to allow the player to follow the avatar from a distance and not taking on the first-person perspective of the avatar herself, the body here still needs to be controlled from outside of the game. Certainly there is a sense in which one is, metaphorically speaking at least, transported into the game such as when one hovers behind one's avatar through the camera body, but still the game world does not envelop every being of the player. The immersion at most occurs when the player experiences a flow that takes place when one is highly engrossed in an activity (Csíkszentmihályi 1996). As all the attention is directed toward the game, there is an experience of being totally immersed by the game environment. However, the actual player outside the game is still indispensable as one who controls what is going on in the game. Although the actual player is wholly engrossed in the game, the environment of the game, as game, still requires the outside player. Thus the most we seem to be able to say is that immersion in the game takes place to a part of the person who is playing the game, and it does not seem possible that the whole of the person can be totally transported and immersed within the game environment.

Perhaps we can solve this problem by going back to the idea of the extended self that we have seen in Chap. 3. According to the view, the boundary of the self is not limited to the skin of the body, but can extend outward as something there can be regarded as part of the identity of the self in question. This is why Otto's notebook is considered a part of Otto himself, as we have seen. Thus the situation where an avatar is intimately connected with the player could at the first sight be seen as a case where the self of the player extends toward the avatar, making the avatar part of the identity of the self of the player herself. This is reinforced by the usual talk about the avatar where the player refers to her avatar by the first-person pronoun. Linderoth (2005) studies how children refer to their avatars in their games by using the pronoun 'I.' When the avatar is hit, for example, the player would say to his friends "*I* am hit." So the identity of the player himself, the sense of who he actually is, covers his avatar too. It can thus be said that the avatar is also an extension of the self of the player. This point makes it more difficult to look at the situation as one where the player is acting out a part. In acting a part the actor becomes the character he is playing. There are no two bodies located separately in two locations, one in the offline space and the other in the game space, as in the case of the player and the avatar. In the latter case, the self of the player extends toward the avatar, thus when the avatar is hit the player would say that he is hit, implying of course that he and the avatar are one and the same. This might give support to the immersion view that Klevjer advocates; however, the difference is that instead of the self of the player being immersed totally within the environment of the game, the Extended Self View would say that there still remain two selves and one is merely extended onto the game space by virtue of the player playing the game. The Extended Self View thus makes it possible to account for the fact that there are always two selves who belong to one and the same person, and it is not the case of one being immersed onto the environment of the other which implies that the former totally becomes the other and thus loses his identity. The avatar is still one and the same as the player—that is what the thesis of the Extended Self View is about—but the one and the same player exists in two manifestations. This way of looking at the situation, I believe, makes it easier get a grip on the complex phenomenon of player-avatar relationship.

This way of looking at things is also tenable in the science fiction case of an avatar existing independently in real space outside of the context of the computer monitor. In the movie *Avatar*, the lead character is a paraplegic marine whose life is confined to a wheelchair. He is hooked up with an avatar system so that he finds himself in a body of an alien race that inhabits the planet that he has been sent to. After the successful hook up, he finds himself 9 ft tall, having a tail, blue skin and so on. So the avatar is the alien creature and while he "plays" the avatar he remains immobile in a tube hooked up with a lot of wires. The avatar and the marine himself, unlike the computer game situation, inhabit the same world. That is, it is always possible for the marine (whose brain directs the avatar) to have the avatar come to the place where he stays and have a look at himself. During the hook up the marine looks at the world through the eye of the avatar, and when the avatar comes to look at how he remains motionless inside the tube, he would be looking at himself as if he himself were an external body, a body of another person and not his own body which now is that of the blue 9 ft tall alien. The *self* of the marine, what Merleau-Ponty calls the phenomenological experience arising through perception based on a body, is now located inside the body of the alien, but the brain that controls everything—including thoughts, emotions, memories and so on—belong to the human who is locked inside the tube. The self of the marine (the thinking part) inhabits the body of the 9 ft tall alien with tail, and it is a real body that can cause real damage to the things in the same world if he is not careful in using the body. Thus when the alien (who is in fact an avatar) speaks of himself, in most cases that would refer to his body that is 9 feet tall. Only in very rare case would the word 'I' refer explicitly to the body of the paraplegic inside the tube. However, that the word 'I' can refer to two different bodies depending on context shows that the self can be extended outside of the skin, and understanding the relationship between the avatar and the person in this science fiction scenario can shed some light on the player-avatar relationship in the game too.

What emerges from this analysis of both the computer game and the movie is that the first-person pronoun can sometimes create confusion as to who exactly is being referred to. This normally does not happen as we usually inhabit only one body. But when we come to inhabit an avatar it has to be made clear to which body the 'I' refers to, which implies that bodies are not all there is to the self or the person. This shows that reference to the first-person pronoun actually depends on the context. When the avatar, the blue alien, uses the word 'I' and 'me,' he has to make sure, unless the context is clear, which one he is talking about. Does the 'I' refer to the blue alien avatar, or the paraplegic marine? For the avatar to say "I am a paraplegic marine" can be either true or false depending on the context. In one sense, the statement is false because the blue alien can run very fast and jump very high, thus he is not a paraplegic at all. In the other sense, the statement is clearly true. This is not supposed to happen because using the first-person pronoun is among the simplest ways of using the language. We should understand why this is the case better if we understood that what constitutes the referent of the 'I' is not a unitary entity occupying a definite chunk of space time, but can be either here or there which can only be so if the self is understood to be a composite, something constructed out of a large number of mental and physical episodes, as we have seen in the previous chapters.

Thus, in the case of the player and her avatar in the computer game, the two are parts of the self of the player. It is the self of the player that extends toward the avatar. This is the reason why it makes sense for the player to say "*I* am hit" when in fact it is his avatar in the game that is hit. There is an interplay and a co-construction of each other in the game. On the one hand, the player constructs his avatar, not only in the sense of choosing what the avatar look like, what gender it should belong and so on, but also the player constructs the identity of his avatar during the course playing the game. This is especially true in online games where the recognition of others plays a big role in building up the identity of an avatar there. Suppose a player assumes an avatar and engages himself in an online game for a certain period of time. Sooner or later the avatar will acquire some reputation among his peer. Perhaps this avatar is very good at a particular style of fighting; thus after a period of time rumor will begin to spread around in the online space that this particular avatar is very good at this style of fight. This spread of reputation plays a large part in building up the *identity* of the avatar in question. This recognition of identity usually results in the avatar getting a name from the community. Suppose the avatar is good at fighting the Kung Fu style; then it is very plausible that the avatar will get the name "Kung Fu" or something of that kind. Here we have a case of offline players building an identity of an online avatar. This building up of identities, on the other hand, can also take place in the other direction. An offline player can get a reputation for being very good at playing a particular game in a distinctive style. Thus it is also conceivable that the player will get a name that denotes his prowess in the online game space too. Hence the identity of the offline person can be constructed by what he does in the online space too. In this case, it is, conversely, the self of the avatar that *extends* toward that of the offline player. The latter becomes known to his circle as the Kung Fu Man (if his fighting style in the online game is Kung Fu). The online world intrudes upon the offline one and makes a change there. And a philosophical import we also get from this is that the use of the first-person pronoun is not a simple affair of using it and it magically and transparently refers to an existing entity that is the speaker of the statement. Context of use is vitally important.

This insight into the composition of the self and how the meaning of indexicals such as 'I' depends crucially on context helps us see how the paradox of the player and the avatar mentioned by Klevjer is resolved. In his chapter Klevjer writes:

> At the heart of the player-avatar relationship lies a tension and a paradox, reflected in our intuitive understanding of what is means to be immersed in a navigable 3D environment through an avatar. How can we say that the player is extending or reaching into the gameworld, while at the same time also saying that the player is "being within" and "acting from within" the gameworld? How can avatarial embodiment be both a kind of *extension* and a kind of *re-location* at the same time? (Klevjer 2012, p. 20).

The paradox can be resolved if it is recognized that context of use plays a necessary role. In one context the player is best regarded as *extending* into the game world. For example, when the separateness between the player and the avatar is emphasized, such as when there are more than one flesh-and-blood players playing one game together through a network, in this case it makes sense to distinguish who is playing which avatar, making it necessary to pair up each player with their own avatar. In another context, however, it is more appropriate to talk about the avatars alone without having to refer back to the real players sitting behind. This kind of talk takes place naturally during the course of the game when things heat up and people are immersed in the game itself. So the paradox is only an apparent one. The Extended Self View appears to do a better job at accounting for this apparent paradox. On the one hand, the avatar and the player are two distinct entities–one is playing the other; on the other hand, the avatar merges completely with the player when the flow of the game directs all the attention to the game itself. This of course does not mean that the real player disappears into thin air when the game is being intensely played, but since all attention is on the action in the game the players forget that they are flesh-and-blood carbon based beings for a moment. This should not be surprising, as we assume different identities all the time depending on changing roles and circumstances. In driving a car, sometimes there arises a situation where it makes sense to use the first-person pronoun to refer not only to the driver herself, but to the whole driver-and-car complex, assuming that there is a merger of the two such that the self of the driver extends onto her car. We can imagine a case where a driver passes another one, saying to the latter, "I am faster than you!" The 'I' here does not refer only to the body of the driver himself, but to the driver-and-car complex which is now moving faster than the complex belonging to the second driver. That it makes sense to say this shows that the first-person pronoun does not always refer to the physical body of the speaker. One's sense of identity can extend to props and machines that lie at one's disposal too, and avatars are clearly one of those props, and the attachment of the person to her avatar is perhaps even stronger than that between the person and her car.

6.3   Fictionalism and the Material Divide

We see, then, that there can be influences that go in either direction, either from games to the real world or the other way round. Someone may gain a reputation in the online world as a particular avatar who is good at one thing, and this reputation can also spill over to the outside world, so the player becomes known to his peer as *this* avatar who is good at this thing. Klevjer, on the other hand, argues that there is a "material divide" (Klevjer 2012, p. 24) that separates the two worlds. This does not imply that the divide is between putatively real (i.e., our own familiar world) and the fictional world one finds in the game. Klevjer contends that fictionality does not have to play a role in accounting for the phenomenon of prosthetic telepresence (Klevjer 2012, p. 29). The game space is one where the player, through the avatar, finds her body-subject, orienting herself as if her real body is there within the space of the game (Klevjer 2012, p. 29). The divide, then, is between the putatively real space in which we all live, and the space of the game. Thus the divide is between how the space is oriented and how one finds oneself as a body-subject rather than between different sets of content (real or fictional) within the space. Klevjer's point, then, is that he seems to want to shift the emphasis from the fictionality of the game space to the different configuration of body orientation as the criterion of the divide. In other words, Klevjer would like to have different forms or configurations of space as what is relevant in the material divide, rather than whether the content is real life on one side and fictional, make-believe life on the other. Fictionality of the game space, in other words, though undoubtedly obtains inside the game, does not function as what separates a game from real life. What does is the fact that the body-subject has to re-orient herself and has to relocate herself within a new set of spatial parameters. However, if we pay attention back to the material content of the game, we still find that what differentiates the world of real life and that of the game in many cases is that the latter world is fictional. My example of a non-existent game (there is actually no Don Quixote game on the market that I know of) based on the story of Don Quixote is a case in point. In any case, that the event in the game may be fictional does not have to preclude the possibility that the two worlds can merge together. The player whose avatar is Don Quixote can gain reputation in the outside world as someone who is very good at playing *as* the Don. An analogy is certainly that of an actor who gains his reputation as someone who plays Hamlet very well. No one is going to believe that the actor is the real Hamlet. In the case of the game player whose avatar is Don Quixote, his identity is thus mixed up between that of the Don in the context of the play (which is not necessarily exactly the same as the Don in Cervantes' novel) and his outside, offline identity. Whatever his latter identity may be, it now has obtained an added dimension as someone who is very good as the Don Quixote avatar. What happens inside the game brushes over to the real world, across the divide if there were one.

However, Espen Aarseth argues that the concept of fictionality does not apply to computer games, and the more appropriate concepts are "virtuality" and "simulation" (Aarseth 2007). The content of fiction—Don Quixote, Smaug the Dragon, etc.—are strictly speaking inside our heads. Our imagination creates them out of the meaning of the words in the literary or dramatic works. Games, on the contrary, are real in that they are the result of the work of the machinery and algorithm of the computing machine. Smaug the Dragon in the game is one we can play with or against, in contrast to Smaug in Tolkien's novels which exist only in the imagination (Aarseth 2007, p. 37). Instead of characterizing elements in the game as fictional, Aarseth argues that they should be more accurately understood as belonging to virtual or simulation environment. The implication is that Smaug inside the game is a virtual dragon, not a real one which does not exist. And this virtual dragon is not there in the novels or in the movies. Thus the virtual dragon is not fictional; it is real in the sense that what is virtual can be real. We could say that Smaug in the game is a real dragon to the extent that the game is real (people can actually run the game on their computers and play it), which makes its elements inside somehow real too. Thus for Aarseth the relevant divide is between what is virtual or simulated and what is really real, in other words what is inside and outside of the game. His divide thus looks quite similar to Klevjer's.

Another difference between games and fiction for Aarseth is that games are interactive. We can play against Smaug in a game, trying to beat in various ways, but when we read the Lord of the Rings saga, we are in no position to change the course of the story. However, there is a genre of game where the player can directly change the course of the story of the game, and this intervention in the story is the essence of the play in the game. *Heavy Rain* is a famous game on Playstation platform that has a moving story line and what is distinctive about this game is that the player is directly involved. They can become one or more of the lead characters in the story and make decisions which will alter the course of the story itself. *Heavy Rain* is full of what Klevjer calls "cut scenes" where movie clips are inserted in a play to give it a story line, and Klevjer argues that they are what make the game fictional (Klevjer 2014). However, in most games cut scenes are put there just to present a story, to give a context to the game and the game itself does not contain those scenes (which led many players to skip cut scenes altogether because they think they are not relevant to the play). In Heavy Rain, on the contrary, the game is totally constituted by 3D movie clips. There is no separation between movie clips and the actual play of the game because the *actual* play is the movie clips themselves. The game starts with a usual movie clip about a typical family, but then an event happens and the player then becomes one of the characters who have to make a decision which will start a chain of events leading from the decision. The game is the story and it is even more so when there is a beginning of the story, a development, and an end. The player cannot choose the beginning, but they can choose the development and interestingly the end. There are several endings in the game depending on how the player has made her decisions in the previous scenarios leading up to the end. Hence the player can in effect choose which ending she likes most. In one ending the lead characters do not die; in others almost all the characters end up dead, and so on. In this game, then, it is rather difficult to say that it is not fictional. The whole game is similar to a kind of genre in novel writing where a number of authors are present and they take turn to write the same novel. One author leads the characters one way and then another take the same characters in the same story another way. The difference is that in *Heavy Rain* several paths of the story line are present at the same time, which is comparable to the situation where the multiple authors of the novel take up several story lines, resulting in there being more than one story consisting of the same basic structure of the story and the same set of characters. Since the "cut scenes" are all there is in *Heavy Rain,* it appears to be as purely fictional as a game can be. Aarseth does not discuss this genre of game in his paper, but according to his argument the story that constitutes the game is more virtual than fictional because we can manipulate the minds and actions of the characters and the direction of the story. As we manipulate the virtual dragon inside another play, we manipulate how the characters react, so perhaps according to his argument he seems forced to conclude that *Heavy Rain* is not a fictional game. However, as there is the practice of multiple authors taking on the same novel as I have just said, and as the novel here is obviously fictional, *Heavy Rain* which follows this genre rather closely should also be considered a fiction too. That the characters in the game also appear on screen and are able to be manipulated should not detract from its fictional character because in the game it is the story line that is the overriding theme of the whole game. A game where one can play a number of characters and influence the direction of the plot still contains fiction. The game itself functions more like a stage in which the fiction unfolds.

Heavy Rain is in fact a special case. It differs from most other games in that it relies on heavy fictional elements. However, there are other games where the fictional element is almost non-existent. A popular game on the iPad and Android devices, *Osmos,* consists of circular blobs, called "motes," varying sizes floating weightlessly in space. The objective of the game is to manipulate a mote so that it eats other smaller motes thereby becoming bigger and avoid being eaten by bigger motes. Thus the player has to move this player-mote around by touching the screen in the direction opposite to the direction to the direction she wants the mote to go. Touching the screen behind the player-mote causes it to eject material from inside its own body, creating a thrust but as it loses its own material the mote becomes smaller as a result and becomes easier to be eaten by other bigger motes. Here the fictional element is almost non-existent. Firstly there is no story at all. There is no narrative, no beginning nor ending. We are not told when we first come to the game where the motes are from and what they actually are. We are told only the objective of the game and how to play it. In this sense motes are motes and do not represent anything outside of themselves. Moreover, the game uses its touch screen platform to let the player navigate the motes and actually play the game. Playing a game by touching the screen tends to reduce any divide that may exist between the game world and the real world ("work" world) because the touch means that the player engages with the game directly within the "work" (real) world. In manipulating game elements using the keyboard, mouse, joystick or buttons on the console, a kind of make believe is there as the controller is hidden from the game world. In a classic first-person shooter game, hitting the keyboard may be a way to control the avatar inside, but the keys in the keyboard do not figure in the game. The keys remain hidden from the context or the setting of the game itself. On the contrary, when we touch the screen in order to move our mote the controlling mechanism interacts directly with the game element. The mote is there right before our eyes within our own "work" world. Since it does not represent anything it stands out as what it really is, a group of pixels on the touch screen of the tablet which we can move around by touching the screen. We can also imagine a technology that allows for three-dimensional motes on a table and we move them by touching the space behind them. In this case the motes belong to the "work" (or "offline" world) and thus there is no gap between the two worlds because there is only one. In this case it seems counterintuitive to say that the motes are parts of a fictional world. They are there in our world, and even when they exist as pixels on the touch screen of the iPad they are parts of our world too. Consequently, there are games in which the elements are highly fictional and those that are minimally fictional at all if at all. Being fictional is thus dependent on the nature of how the elements within the game are representational or not, or whether they are intended by the designer to tell a story or not.

Furthermore, the possibility offered by my fictional game discussed earlier that I could become Don Quixote is what is missing in Klevjer's phenomenological analysis. What his analysis offers is an intricate account of how my body, either in its physical form or its digital form as it finds itself inside the game, is oriented in relation to other props in the game so that I find myself having a definite location within that space. The emphasis on the relations and the form of being situated inside a space is thus too general to account for the very specific fact that I become Don Quixote in this game, in this very space offered by the game. I certainly enter into a different set of configurations on my body-subject, but these configurations say nothing about my being Don Quixote and not, say, Julius Caesar. Suppose that all configurations of the space when I am Don Quixote and when I am Julius Caesar are absolutely the same. In the two spaces I find myself oriented as up, down, left, right in exactly the same way, but in the first space I am Don Quixote and in the second I am Julius Caesar, there would be nothing in these parameters to account for this difference. Hence, the role of fictionality in separating what Walton calls the "game world" from the "work world" (Sageng et al. 2012, pp. 180–181) is still relevant. Furthermore, Grant Tavinor (2012), and Aaron Meskin and Jon Robson (2012,) argue that the concept of fiction still works as a tool to describe and help one to understand the phenomenon of the computer game. According to Walton, works of fiction are objects that "serve as props in a game of make believe" (Walton 1990, quoted in Sageng et al. 2012, p. 180). Instead of trying to analyze the language of fiction and solve the puzzle of how the semantics of the language succeeds in referring to fictional objects, Walton looks at how the language achieves its aim in a pragmatic manner. That is, how the language succeeds in making us believe that there is a setting in the fictional world where there are such and such characters doing such and such things. We as readers know that these are not real, but we willingly enter into what Walton calls the game of make-believe, suspending our judgment of what is real for a moment when we enter the world of the fiction. In this sense works of fiction serve as "props" that help us construct a game of make-believe. Thus Tavinor and Meskin and Robson present their arguments purporting to show that computer games could also be considered as such a prop that help us enter the world of make believe too.

If there is a difference between the game world and the work world (which, by the way, is similar to Klevjer's material divide that we have seen earlier), then what about the self of the player who enters the game world when she plays the game and goes back to the work world when she quits? I have argued in the previous chapters that the world of the online personae and the offline world are not strictly separated, but merge together in many ways. Is the divide between the game world and the work world (the latter of course means that it is not a "game" where one "plays") the same as the divide between the online and offline worlds that we have seen in the previous chapters? If we look at the context of game playing then a difference is that in playing the computer games, one does not enter into the game always as oneself, but one assumes an identity of an avatar which could be wildly different from what one is in real life. But certainly in entering the online arena one can also dress up oneself to be very different from what one is too. This is a familiar phenomenon. The difference, however, is that in the online setting where, such as in social networking sites, one dresses up one's profile and even presents oneself anonymously, the situation one finds oneself in (that is, inside the cyberreality of the social networking site) is taken to be a real one whereas what happens in a game is fictional. When I assume the avatar of Don Quixote or the mustachioed Mario I enter into a fictional world; even in the setting of online games such as *League of Legends* I still enter a fictional world where I become a medieval warrior. But the world inside the social networking site is not fictional. It is intended to be a part of the "real world" or in other words "work world" (where "work" means "not a game"). The very context of a game—it is where one *plays*—makes it the case that what is happening there is not supposed to have any bearing on what is happening outside. In the social networking, on the contrary, this can always happen, as people can make appointments inside the space of the networking site in order to meet up in real life afterwards, for example.

6.4 Brain to Brain Integration and Complete Erasure of the Boundary

Another topic which has gained some traction in the literature (see, for example, Hongladarom 2015) is the possibility of brain-to-brain integration, where the brains of more than one person are linked up through networking so that information passes through directly from one brain to the other without relying on external medium. In other words, two brains are linked as if they were to become one brain, where thinking passes through instantaneously from one brain to another without relying on speech, for example. This is a natural development of the research and development on brain to computer integration, where the brain is linked up with a computer directly so that information from the brain can feed directly to the computer, enabling the person owning the brain to issue commands to the computer with her thoughts alone. This already has a number of applications enabling disabled patients, for example, to communicate with the outside world and to do things through the computer in a way that was not possible before. In the case of brain-to-brain integration, then, the brains are linked up in the same way, which opens up many possibilities. (For more information on the topic, see Hongladarom 2015).

One of the possibilities that I would like to discuss in this section is that with such close integration of brains there is a possibility that the boundary between what is taken traditionally to be the self and whatever exists outside the self can be completely erased, or at least there is a possibility that such a boundary will not be a hard and fast one, but something conventionally located, such as a line separating different lanes on a highway. (Paul Verbeek also analyzes the boundary between human and technology in Verbeek 2009.) This possibility has strong potentials for gaming and for reflecting about the lines between games and the outside world. Firstly, in the context of team playing, we could imagine a game between two teams, each consisting of two or more persons with their brains linked up together, or the team here might also compete against the computer. The idea is that, instead of working together as distinct individuals, the team members, with their brains connected in this way, can function as *one* unit, thus erasing the time lapse needed for verbal communication. Even non-verbal communication that usually exists among team members, such as in tennis double matches, will be superseded when the brains of the two players are connected in this way. In a sense, then, the two persons with their brains connected directly could become one person; indeed how to count them as one or two persons would depend more on what we are counting rather than something that objectively exists on the outside. On the one hand, there are obviously two bodies, but when the brains are merged in this way, there is a sense in which they have become one. I have discussed some of the ethical conundrums that emerge from such an integration elsewhere (Hongladarom 2015). Here I would like to discuss more on the implications that brain-to-brain integration has on computer games and online selves in general, and the first point that I have been discussing so far is that the boundary between one person and another will be more a matter of convention rather than something that exists objectively. If this is the case, then the boundary between the self of the player and the avatar can also be considered in the same vein, namely that there is no objective boundary between the self of the player and the avatar either.

Secondly, when persons and their brains are merged in this way, we can imagine a game where there is an avatar controlled by a team of brain-connected players. This avatar can then perform all the tasks and compete with other avatars which presumably are run by brain-connected teams also. There are a lot of conceptual problems in this scenario that need to be unpacked in order for us to understand what is exactly going on and so that we can anticipate the situation when it eventually arrive. Here the focus is not the conceptual ramifications of the relation between the player and the avatar as we have seen in the earlier section. The focus is instead of how it is possible for two or more human players to inhabit one and the same avatar, especially with their brains linked together. When one player decides to move an arm of the avatar, does the other have to concur? Or will there be a domination of one player by the other so that the former will function just like a set of limbs of the latter only? All these are very important questions, but they cannot be answered satisfactorily in this section in a chapter on online selves and computer games. In this section we will consider only the situation where there is a cooperation between the players who have their brains connected but without one dominating the other. Domination in this sense is an ethical problem, which requires its own separate discussion. In case where there is cooperation (in the same way as team members cooperate with each other in a game, such as in tennis double matches), the players join forces and the ideal situation would be that any decision made by the team will be the one made jointly by the team members themselves without any conflict. In fact in some game situation that requires fast thinking there is no time for the team members to discuss about the best move. Any decision has to occur very fast. In this case, it might be conceivable that the decision is made by the two brains thinking together and coming up with the best move. We can imagine this better if we imagined that the two brains in this situation actually become one larger brain.

This situation is admittedly very far-fetched and it will be some time in the future before this becomes a reality. Nonetheless, the point that concerns us at the moment is the *conceptual* ramifications of the situation, especially with regards to metaphysics. Here games are an extension of reality; there does not seem to be an objective boundary separating the two. In the situation where a team of brain-connected players instantiate a common avatar, we have essentially a three-way relation. This becomes more interesting when the players are immersed in the game environment. Here we have two persons merging with each other through their connected brains, and the merged player in a sense becomes a new entity consisting of two bodies of the players and two brains (which are joined together). The emerging player thus finds herself immersed in the game environment, instantiating an avatar. The metaphysical problem is thus how one is to tell how many persons there are inside the avatar. On the one hand, the avatar is controlled and is embodied by the emerging player, but then the latter consists of two bodies linked up through network connection. At any rate, the lesson that results from this is that the boundary between persons themselves are not hard and fast. We seem to believe traditionally that persons are obviously and objectively distinct one from another. The self of a person stops at her skin, but, in addition to the argument for the Extended Self View that we have seen, we are seeing here also that in the context of brain-to-brain integration and gaming, the self of a person is able to extend to another person creating a new, merged identity, as well as to external objects, such as the game environment through the avatar. The *physical* possibility of brain-to-brain connection and integration points to the *conceptual* possibility of porous personal boundaries.

References

Aarseth, E. (2007). Doors and perception: fiction vs. simulation in games. *Intermédialités: histoire et théorie des arts, des lettres et des techniques* [Intermediality: History and Theory of the Arts, Literature and Technologies], *9*, 35–44. URI: http://id.erudit.org/iderudit/1005528ar

Csíkszentmihályi, M. (1996). *Creativity: Flow and the psychology of discovery and invention.* New York: Harper Perennial.

Fine, G. A. (1983). *Shared fantasy.* Chicago: University of Chicago.

Hongladarom, S. (2015). Brain-to-brain integration: Metaphysical and ethical implications. *Journal of Information, Communication and Ethics in Society, 13*, 205–217.

Klevjer, R. (2006). *What is the avatar?: Fiction and embodiment in avatar-based singleplayer computer games.* Unpublished Ph.D. dissertation, University of Bergen.

Klevjer, R. (2012). Enter the avatar: The phenomenology of prosthetic telepresence in computer games. In J. R. Sageng, H. Fossheim, & T. M. Larsen (Eds.), *The philosophy of computer games* (pp. 17–38). Dordrecht: Springer.

Klevjer, R. (2014). *In defense of cutscenes.* Retrieved from http://folk.uib.no/smkrk/docs/klevjerpaper.htm

Linderoth, J. (2005). *Animated game pieces: Avatars as roles, tools and props.* Paper presented at the Aesthetics of Play conference in Bergen, Norway, 14–15 October 2005. Retrieved from http://www.aestheticsofplay.org/linderoth.php

Merleau-Ponty, M. (1962). *Phenomenology of perception* (C. Smith, Trans.). London: Routledge.

Meskin, A., & Robson, J. (2012). Fiction and fictional worlds in videogames. In J. R. Sageng, H. Fossheim, & T. M. Larsen (Eds.), *The philosophy of computer games* (pp. 201–217). Dordrecht: Springer.

Sageng, J. R., Fossheim, H., & Larsen, T. M. (Eds.). (2012). *The philosophy of computer games.* Dordrecht: Springer.

Tavinor, G. (2012). Videogames and fictionalism. In J. R. Sageng, H. Fossheim, & T. M. Larsen (Eds.), *The philosophy of computer games* (pp. 185–199). Dordrecht: Springer.

Verbeek, P. (2009). Ambient intelligence and persuasive technology: The blurring boundaries between human and technology. *NanoEthics, 3*(3), 231–242.

Walton, K. (1990). *Mimesis as make-believe: On the foundation of the representational arts.* Cambridge, MA: Harvard University Press.

Conclusion

In this brief concluding chapter I would like to suggest some directions in which future research on the online self could develop. As we have seen in the book the self is a highly complex and multifaceted concept, and as the self has found its presence in the online world, the complexity is more than doubled because there is the added dimension afforded by the cyberworld with its own multifaceted complexities in many levels. Perhaps, philosophically speaking, the most important direction for future research lies in the argument for or against the Extended Self View that I propose in the book. Does it actually make sense to talk of the self as extending toward inanimate object such as the computer screen? As the argument in the third chapter shows, I believe it is, but then, as with other philosophical proposals, this invites further discussions and debates on the issue, which will only deepen our own understanding of the self, either in the offline or online versions. I have not touched upon the closely related topic of consciousness much in the book, but it is tempting to investigate where the consciousness is going to be located if the Extended Self View is accepted. This is where I think the discussion in the last chapter on computer game becomes relevant. We can seriously talk of ourselves being located inside the context of a game, through an avatar. In this case the location of our selves lie not quite inside our bodies as the players, but inside as avatars who are doing their own things in the context of the game. Thus the question is: Does this imply that the location of our selves, our consciousness, lies outside of our bodies and inside the avatar? Is such a talk like this merely metaphorical, or does it have some germ of truth? These are fascinating questions. I have provided a sketch of answer according to my own proposal in Chap. 6 and also in Chap. 3, but this only invites further research, especially on the boundary between the person of the player and the game, and between the avatar and the person behind it.

Secondly, as we have seen in Chap. 2, the historical account of the concept of the self is a very rich strand of stories with many similarities and differences. One thing future researchers might want to take up is to look at the issue in a more historical manner. Perhaps she might become interested in looking closely at Spinoza's idea on the self and see how that helps us understand the online self better. The view of

© Springer International Publishing Switzerland 2016
S. Hongladarom, *The Online Self*, Philosophy of Engineering and Technology 25,
DOI 10.1007/978-3-319-39075-8

the self in Buddhism is also very interesting, and certainly merits close study which will illuminate not only our understanding of the self, but other metaphysical notions too, as well as the problems surrounding the boundary between persons and between persons and external objects as we have seen. Buddhist philosophy clearly argues that the self is only taken to exist ultimately only by mistake, in the same way as the rainbow, whose objective existence is not there (what are there are perhaps only water droplets and light), but nonetheless perceived to be there and can be talked about as if it were actually existing. But if this were to be the case, then the online self is also conventionally taken to be there in the same way, which implies, for one thing, that the line between the offline and online selves do not exist objectively. I argue that this Buddhist position helps us understand many of the complexities of the issues better than its rivals, but then this again invites further discussion and research. Then in Chap. 4 I outline what traditional philosophy of technology might have to say about the online self. The content of this chapter is very much a sketch, as the online self is not a popular topic in the field yet. Nonetheless, the critical perspectives afforded by the tools of philosophy of technology can yield illuminating insights into how the online self is related to the wider socio-cultural contexts.

The chapter on online friends also provides ample avenues for further research. I have argued that the highest form of friendship, such as the one described by Aristotle, does not in principle preclude online friends. Here the online selves interact with one another mostly through the social media, and I have argued that such interaction could lead to a form of friendship that is genuine and beneficial to each other. This is based ultimately on my earlier argument that there is no real separation between the offline and the online worlds. If the highest form of the good does exist in one world, then it is very likely to exist in the other world too. Further research certainly is needed for empirical investigation of my philosophical speculation here. This would be very interesting and beneficial. Perhaps a way empirically to measure the level of friendship (whether the friendship is at a low level, or higher in the Aristotelian sense, for example) could be developed, and this would be a great accomplishment in itself. And once such a tool is developed, it can then be used to measure the level of friendship that exists both in the offline and online worlds. If it is found that the level of quality of friendship is consistently lower in the online world when compared with the offline world, then one could interpret the situation in such a way that the situation here still needs some way to go before it can arrive at the ideal. The fact that one cannot find an actual situation where online friends achieve the highest level of quality does not show that it is impossible to do so. Alternatively, one might take another route and argue that, since empirical research shows that online friendship cannot achieve the same level of quality as its offline counterpart, then online friendship must be inherently inferior. Either way this would be an exciting avenue of research and discussion.

The last chapter focuses on the role of online selves in computer games. Here I have mainly analyzed the situation and followed the main literature in the field. I also argued that the Extended Self View appears to do a better job at explicating some difficult conceptual issues that are involved, such as the relation between the avatar and the player. I also include a section on a new development in scientific research where information is directly shared between brains. The brain-to-brain integration technology holds a lot of potential for the future. Here the discussion needs to be speculative, as the technology is still at a very early stage. Research on this topic could include conceptual analysis of what it is to be a person—if brains are merged in this way, then does it mean that there emerges a new, superperson consisting of two brains, or one enlarged brain arising from merging the original two? The question has strong ethical dimensions too, though I did not take this up in the chapter. Nonetheless that would be a fascinating topic for research in the near future.

Index

© Springer International Publishing Switzerland 2016   169
S. Hongladarom, *The Online Self*, Philosophy of Engineering and Technology 25,
DOI 10.1007/978-3-319-39075-8